AF346173

Bouncing Signals: A Guide to Skywave Propagation and Ionospheric Sounding

Shivani

Copyright © 2024 by Shivani

All rights reserved. No part of this book may be reproduced in any manner whatso-ever without written permission except in the case of brief quotations
embodied in critical articles and reviews.

First Printing, 2024

Contents

1 Introduction   **1**
1.1 Background to Project 1
1.2 Project Description 3
1.3 Problem Formulation and Research Questions 4
1.4 Scope and Limitations 4
1.5 Plan of Development 5

2 High Frequency Ionospheric Propagation   **6**
2.1 Properties and Structure of the Ionosphere 6
     2.1.1 Ionospheric Layers 6
     2.1.2 Regular Ionospheric Variations 7
     2.1.3 Irregular Ionospheric Variations 9
     2.1.4 The High-Latitude Ionosphere 9
     2.1.5 Ionospheric Models 10
2.2 Magnetoionic Theory 11
     2.2.1 The Appleton-Hartree Dispersion Relations 11
     2.2.2 Simplified Ray Theory 13
2.3 Ionospheric Absorption of HF Radio Waves 15
     2.3.1 Non-Deviative D/E Layer Absorption 16
2.4 High-Latitude Absorption Events 17
     2.4.1 Polar Cap Absorption Events 17
     2.4.2 Auroral Radio Absorption 18
2.5 HF Propagation Prediction Models 19
2.6 Chapter Summary 21

3 HF Antenna Considerations   **22**
3.1 Basic Antenna Parameters 22
     3.1.1 Antenna Efficiency, Directivity and Gain 22
     3.1.2 Take-off Angle and Beamwidth 22
     3.1.3 Front-to-Back Ratio 23
     3.1.4 Bandwidth 24
     3.1.5 Antenna Feed and Impedance Matching 24
3.2 Antenna Polarisation for HF Skywave Applications 25
     3.2.1 Vertical vs. Horizontal Polarisation 25
3.3 The Impact of the Surrounding Environment 26
     3.3.1 Investigation into the South Pole Ground 26
3.4 Suitable Antenna Conductor Materials 27
3.5 Brief Overview of Directional HF Antennas 28
3.6 Long Wire Travelling Wave Antennas 29
     3.6.1 Terminated V Antenna 30
     3.6.2 Terminated Rhombic Antenna 31
3.7 Yagi-Uda Antennas 32
     3.7.1 Horizontal Dipole Yagi-Uda 32
     3.7.2 Resonant Loop Yagi-Uda 33
3.8 Chapter Summary 36

4 Ionospheric Path Analysis and Link Budget   **37**
  4.1 Path Availability Analysis 37
      4.1.1 Results 38
      4.1.2 Discussion 38
  4.2 Elevation Angle Analysis 40
      4.2.1 Results 40
      4.2.2 Discussion 40
  4.3 Transmission Loss 42
  4.4 Ionospheric Absorption Predicted by ICEPAC 42
      4.4.1 Results 42
      4.4.2 Discussion 43
  4.5 Manual Non-Deviative Absorption Calculations 43
      4.5.1 Method and Assumptions 43
      4.5.2 Results 46
      4.5.3 Discussion 47
      4.5.4 Final Ionospheric Absorption Loss Estimates 49
  4.6 Receiver Properties 50
      4.6.1 Receiver Sensitivity 50
      4.6.2 Receiver Antenna 50
  4.7 System Losses 51
  4.8 Link Budget Summary 51
      4.8.1 Discussion 52
  4.9 Chapter Summary 52

5 Antenna Design and Simulation   **53**
  5.1 Design Requirements 53
      5.1.1 Technical Specifications 53
      5.1.2 Simulation Optimisation Goal 53
  5.2 FEKO Simulation Model and Radiation Pattern Plots 53
  5.3 Ground Plane Model Considerations 55
      5.3.1 Construction of South Pole Ground Dielectric Model 55
      5.3.2 Results of Using Different Ground Plane Models 55
      5.3.3 Discussion 56
  5.4 Terminated Sloping V Antenna Design 57
      5.4.1 Choice of V Antenna over Rhombic Antenna 57
      5.4.2 Physical Constraints and Simulation Conditions 58
      5.4.3 Design Methodology 58
      5.4.4 Simulations and Results 59
      5.4.5 Discussion 61
  5.5 Horizontal Dipole Yagi-Uda Antenna Design 61
      5.5.1 Choice of Yagi-Uda Configurations 61
      5.5.2 Physical Constraints and Simulation Conditions 62
      5.5.3 Design Methodology 62
      5.5.4 Simulations and Results 63
      5.5.5 Discussion 64
  5.6 Resonant Loop Yagi-Uda Antenna Design 66
      5.6.1 Choice of Loop Shapes 66
      5.6.2 Physical Constraints and Simulation Constraints 66
      5.6.3 Design Methodology 66

5.6.4 Simulations and Results 67

5.6.5 Discussion 68

5.7 Summary of Antenna Designs Considered 70

6 Final Antenna Design and Analysis **72**

6.1 Final Considerations 72

6.1.1 Antenna Resonance 72

6.1.2 Ground Plane 72

6.2 The Effect of Wire Element Diameter 73

6.2.1 Results 73

6.2.2 Discussion 74

6.3 Final Antenna Design for Practical Element Material 75

6.3.1 Results 75

6.3.2 Discussion 76

6.4 Antenna Feed and Matching 78

6.5 Support Structure and Effect on Performance 79

6.5.1 Results 80

6.5.2 Discussion 80

6.6 Impact of Antenna Height 80

6.6.1 Results 81

6.6.2 Discussion 82

6.7 Robustness of Final Antenna Dimensions 83

6.7.1 Results 83

6.7.2 Discussion 85

6.8 Final Antenna Summary 86

6.8.1 Dimensions and Layout 86

6.8.2 Alignment with Technical Specifications 86

7 Conclusions **89**

7.1 Answered Research Questions 89

8 Recommendations **92**

References **93**

A HF Link Analysis Auxiliary Material **97**

A.1 VT Ray Tracing Tool Figures 97

A.2 Additional ICEPAC Data 98

A.3 Manual Non-Deviative Absorption Calculations 99

B HF Antenna Design Auxiliary Material. **100**

B.1 South Pole Ground Plane 100

B.2 FEKO Radiation Pattern Examples 100

B.3 Additional Preliminary Antenna Design Radiation Patterns 101

B.4 Additional Final Antenna Radiation Patterns 102

B.5 Impedance Matching Equipment 102

C Supporting Literature Auxiliary Material **103**

1 Introduction

The ionosphere is a section of the earth's upper atmosphere in which ionisation by solar radiation occurs. This results in a large concentration of ions and free electrons in the region, which increases with altitude before peaking and tapering back down again. A radio wave propagating through the ionosphere will excite any free electrons present, causing them to re-radiate the wave. This results in the gradual refraction of the wave away from the region of increasing electron density. Consequently, some radio waves can be "bounced" off from the ionosphere multiple times. This is known as skywave propagation and allows for over-the-horizon communications and ionospheric sounding. The high frequency (HF) band ranging from 3 MHz to 30 MHz is most effectively used for skywave propagation, as lower frequencies tend to be heavily absorbed, whilst higher frequencies are not bent back towards the earth sufficiently and simply continue to propagate through into space.

The South African National Space Agency (SANSA) intends to implement a bistatic radar ionospheric sounding system between the South Pole and the South African National Antarctic Expedition (SANAE) base station. HF continuous waves are to be transmitted from a low-powered beacon installed near the geographic South Pole and received by the SuperDARN radar situated some 2090 km away at SANAE station. The primary objective of this system will be to detect and characterise travelling ionospheric disturbances (TIDs) in the polar regions. This book concerns all aspects of the design of the HF transmitter antenna most suitable for successful system operations.

1.1 Background to Project

Atmospheric gravity waves (AGWs), as the name suggests, are buoyancy waves that occur in the atmosphere and produce travelling fluctuations in the density of atmospheric gas. These AGWs are an important means of energy exchange between the neutral and charged particles in atmospheric plasma and can propagate for thousands of kilometres away from their original source. The causes of such waves are numerous and can be related to space-weather events (e.g. particle precipitations and Joule heating) or driven by processes occurring near the surface of the earth (e.g. earthquakes and convective thunderstorms). Knowledge and tracking of AGWs helps to identify the initiating sources and better understand the coupling of energy between the vertical and horizontal layers of the earth's upper atmosphere. AGWs occurring at mid-latitudes, as well as near the auroral oval, have already been studied and modelled fairly in depth. On the other hand, in polar cap regions (particularly in the Southern Hemisphere) observational data is in short supply [1]. The polar cap regions are particularly susceptible to AGWs driven by space-related phenomena because of the geomagnetic field lines connecting them to the magnetosphere [2]. Observing such AGWs would be a worthwhile investigation in improving knowledge of the effects of space-weather events on the earth's climate.

TIDs are simply indicators of AGWs and manifest as quasiperiodic fluctuations in the total electron content (TEC) of the ionosphere. These changes in TEC have an effect on the maximum usable frequency (MUF) and other parameters of the radio wave needed for various communication systems (e.g. HF radio skywave links and satellite communications) and can consequently result in signal being lost. However, these same radio signal modulations are what allow for the use of remote sensing techniques to detect and characterise TIDs, which is the aim of the system under consideration in this book.

The Super Dual Auroral Radar Network (SuperDARN) consists of a set of 35 low-power HF radars placed around the globe. The majority of these radars are located at high-latitudes in the Northern Hemisphere, with only nine active in the Southern Hemisphere. The SuperDARN radar installed at the SANAE base consists of a recently upgraded (2013) fully digital transceiver with an antenna array of horizontally polarised twin-terminated folded-dipoles (TTFDs) designed for operation over the 9 MHz to 17 MHz frequency band. The SuperDARN radars were chiefly developed for measuring ionospheric plasma convection by coherently processing back scatter received off magnetic-field aligned irregularities in the ionosphere. The total received back scatter can contain elements of both ionospheric and ground scatter for up to 1.5 ray hops [3].

The general properties of TIDs can be deduced from perturbations in the amount of ground scatter signal received by a SuperDARN radar. However, this method of detection is not the best for studying TIDs in detail because it does not actually directly measure any of the TID parameters [4]. An alternative and improved method would be to use the frequency and angular sounding (FAS) technique developed by [5], combined with the TID modelled as segments of perfectly reflective surfaces. This technique uses measurements of Doppler frequency shift, propagation delay and angle of arrival (AOA) over a bistatic HF skywave communication link in order to acquire the fundamental TID parameters (amplitude, wavelength, direction and velocity). A simple schematic of the this can be seen in Figure 1.1. Both the azimuth (xy-plane) and elevation (xz-plane) angle of arrival would need to be determined.

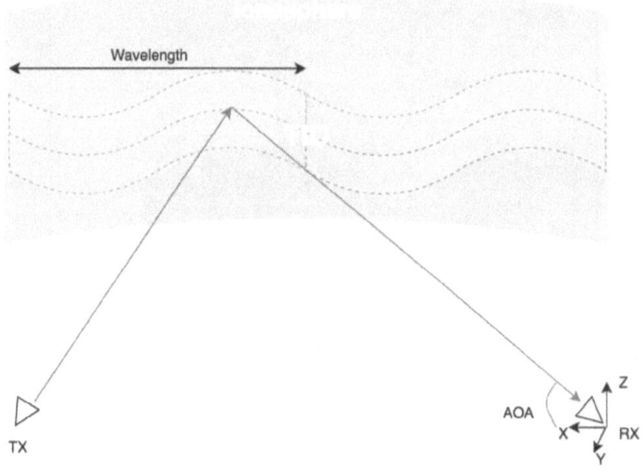

Figure 1.1: Detecting travelling ionospheric disturbances with a bistatic radar configuration.

1.2   Project Description

A transmitter system needs to be installed at the geographic South Pole in order to implement the type of bistatic ionospheric sounding system discussed above. SANSA has already designed and tested a transmitter beacon, however, a suitable transmitter antenna has yet to be procured. Problems arise because there is already an existing SuperDARN radar situated at the South Pole base station. Therefore, permission to install a transmitter nearby is granted on the basis that as little radiation as possible will be directed towards the South Pole SuperDARN so as to minimise interference with it. In order to meet this specification, the design of a highly directional antenna is compulsory. That is to say, an antenna with low backlobe radiation is required. Additionally, SANSA gave the following user requirements for the transmitter antenna:

1. Horizontal polarisation and single operational frequency at 12.57 MHz.

2. A maximum power supply of 1 W. The aim, however, is to transmit as little power as possible. A minimum of 100 mW is allowed by the transmitter beacon.

3. A maximum voltage standing wave ratio (VSWR) of 1.5. Any matching techniques required should be explored.

It would also be beneficial if the designed antenna simultaneously allowed for the illumination of one of the two SuperDARN radars nearby to SANAE IV, namely Syowa Station and Halley Station. This is not essential, but would be a useful feature to incorporate into the system, should the need to cover additional areas arise. A map of the respective stations, their fields of view and their angular distance from SANAE relative to the South Pole can be seen in Figure 1.2. Antenna radiation pattern specifics are expanded on in the later chapters of this report.

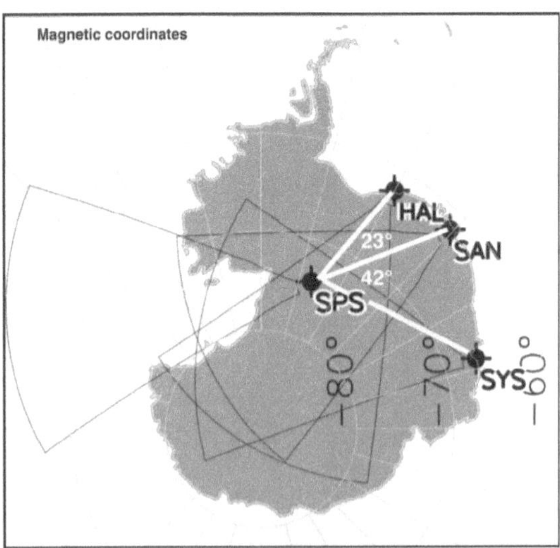

Figure 1.2: Map showing Halley (HAL) and Syowa (SYS) SuperDARN radars and their angular distance from SANAE (SAN) relative to the South Pole Station (SPS).

1.3   Problem Formulation and Research Questions

The aim of this research is to design a highly directional low-powered transmitter HF antenna for installation at the South Pole. Before any actual antenna design can take place, it is vital to first establish what antenna radiation profile will most effectively allow for communication over the path. This leads to the principle research question:

> *1. What transmitter antenna gain, power and range of elevation take-off angles is needed to achieve successful communication with the SANAE SuperDARN receiver?*

To answer this, an in depth account of the most notable losses seen by the radio wave on its expected path is needed. This prompts the next set of research questions:

> *2. What suitable ionospheric models and ray tracing methods are easily available for propagation predictions? How valid are these models when applied to the polar regions? What significant shortcomings do they have and how could these potentially be overcome?*

> *3. At what times of day and on which dates will the system most likely be able to consistently operate?*

> *4. What is the biggest contributor to ionospheric path loss? What is the peak ionospheric loss that may be experienced by the transmitted signal in standard ionospheric conditions? At what time, season and stage of the solar cycle will this peak occur?*

Furthermore, in relation to the environment in which the antenna will exist:

> *5. What are the dielectric properties of the ground plane at the South Pole? What effect will this have on the antenna performance?*

Ultimately, the aim is to answer:

> *6. What antenna design will meet the developed requirements to the highest degree, whilst being simple to construct, inexpensive and able to withstand the South Pole environment?*

1.4   Scope and Limitations

The designed antenna will have to be transported to the South Pole where it will be assembled on site and left for extended periods of time. The equipment available for constructing the antenna will be limited, so erecting very high antenna masts will be particularly difficult. The height to which the antenna extends will therefore need to be as truncated as possible. The span of the antenna in the horizontal plane is not really an issue. However, any of the structures installed should be easy to remove once the experiment comes to a close so as to leave no trace behind. The materials used will also have to be restricted to those that are able to withstand extreme cold, winds, snow and significant UV exposure.

The ionosphere is a highly variable medium and it is therefore unlikely that the system will be operational all year round. System optimisation is thus aimed at periods when the ray is anticipated to reach the receiver most often. Accordingly, there is a heavy reliance on propagation models which are not necessarily accurate for use in the high-latitude ionosphere. Furthermore, the performance of the antenna in a simulated environment will be unlikely to match that of its real world performance, even in the best cases. Therefore, building and testing of the antenna would certainly need to be done in order to accurately rate the performance of the designed antenna, but was not included in the scope of this **book**. That being said, testing the antenna in South Africa would not necessarily be a good indicator of the performance that would be obtained at the South Pole due to the dissimilar environment. A satisfactory antenna can still be designed with simulation alone, as long as sufficient safety margins are included and care is taken when modelling the nearby environment.

Ensuring that transmitted radio signals will actually reach the receiver with enough power is the part of the bistatic ionospheric sounding system on which the research presented in this **book** is primarily focused. Any signal processing that would need to be done at the receiving SuperDARN to extract TID information will not be considered.

1.5   Plan of Development

A brief outline of the remaining chapters is given here:

- Chapter 2 contains the theory and literature review of HF ionospheric propagation. It commences by introducing the ionosphere and some of its properties at high-latitudes. Models of the ionosphere and other geophysical properties are also reviewed. Magnetoionic theory, simple ray tracing methods and available propagation prediction tools are critiqued in depth so as to enable a thorough link budget analysis to be conducted.

- Chapter 3 contains the theory and literature review of HF antenna design. Relevant HF skywave antenna design parameters are presented and the specific dielectric properties of the ground plane at the South Pole are investigated. A review of the most suitable antenna candidates and the theory behind their operation is undertaken.

- Chapter 4 makes use of the findings in Chapter 2 in order to establish the criteria needed to be met by the designed antenna. Selected available prediction tools are used for the analysis of system path availability and required ray take-o angles. The applicable magnetoionic theory is implemented in order to estimate a portion of the most notable wave absorption loss. This is used in combination with the losses predicted by propagation prediction software. A link budget analysis is then done in order to determine what transmitter antenna gain and power is required.

- Chapter 5 deals with preliminary antenna designs and simulations. It details the design approach taken for each chosen antenna and the results of their subsequent performance. Based on the findings of this chapter, the most appropriate antenna was selected for the final design.

- Chapter 6 deals with all aspects of the optimised final antenna implementation and gives a summary of the performance achieved through simulations.

- Chapter 7 concludes the work presented in this **book** by answering the set of research questions posed.

- Chapter 8 provides follow-up research recommendations.

2 High Frequency Ionospheric Propagation

The first part of this chapter examines the structure of the ionosphere and its impact on propagating HF waves, with a focus on high-latitudes. An evaluation of a few available ionospheric models and their validity in the remote polar regions was undertaken. The second part of this chapter discusses the basic principles behind ionospheric propagation modelling and a common ray tracing method used to predict the HF channel. An important aspect of this section is the formulations for the non-deviative absorption that occurs in the lower layers of the ionosphere. These were used in the link budget analysis to quantify some of the regular ionospheric loss that a radio wave may encounter. Specific high-latitude absorption events and their effect on system operations were also considered. Finally, a few of the freely available HF propagation tools were researched and selected for use in the ionospheric path analysis and link budget.

2.1 Properties and Structure of the Ionosphere

The ionosphere is generally accepted to extend from 50 km to 1000 km above the surface of the earth. It is composed of positively charged ions and electrons which are formed in the neutral atmosphere when solar radiation separates the electrons from atoms and molecules. The process is known as photoionisation and can be described by,

$$A + hv \rightarrow A^+ + e^- \tag{2.1}$$

where hv refers to the photon flux (the product of Planck's constant h and photonic frequency v) and A refers to the neutral atom or molecule. The dominant species of ions in the lower half of the ionosphere are molecular Oxygen (O_2^+), Nitrogen (N_2^+) and Nitrogen Oxide(NO^+). Atomic Oxygen (O^+) ions become prevalent in the upper half of the ionosphere, alongside a small percentage of atomic Nitrogen (N^+), Helium (He^+) and Hydrogen (H^+).

The level of ionisation that occurs will depend on the radiation intensity and the surrounding neutral atmosphere composition and density. As altitude increases, radiation intensity also increases while neutral atmospheric density decreases. Peak ionisation levels will occur in regions where both the surrounding neutral density and the radiation intensity are sufficiently high. When nightfall comes, photoionisation will largely cease and the electrons and ions will begin to recombine to produce neutral atoms. At night, ionisation levels are significantly reduced but never completely depleted. This is especially true in the upper ionospheric regions, where the ionisation levels will ordinarily remain high enough to still support HF radio sounding [6].

2.1.1 Ionospheric Layers

The ionosphere can be divided into several principle layers each with different characteristics. The layers, namely D, E, F1 and F2, are typically defined according to their electron density profile. A minimum inflection in the electron density profile often identifies the boundaries of the layers, although this can frequently be indistinct [7].

The approximate height profile, electron densities and classical role in HF communications of the principle layers are discussed below. The exact layer height and electron density definitions vary from source to source and were taken here from [7] with respect to a mid-latitude ionosphere.

D Layer [60 km - 90 km]: Electron density $10^8 - 10^{10}$ m^{-3}

This is the lowest classical layer. It is characterised by the highest surrounding neutral atmosphere density of all the layers, making it responsible for the majority of HF wave ionospheric absorption. This is mostly owing to electron-neutral collisions. The D-layer can sometime extend down to a 50 km altitude, which is a segment referred to as the "C layer" and accounts for additional ion production from galactic cosmic ray contributions. The D Layer tends to disappear at night when ionising solar radiation is no longer present.

E Layer [105 km - 160 km]: Electron density 10^{11} m^{-3}

This layer of the ionosphere discharges more slowly than the D-layer and, although weakened, is still present at night. It also contributes towards some radio wave ionospheric absorption. Sections of higher electron density called sporadic E (E_s) can occur as a result of pressure changes. This anomaly is linked to solar and geomagnetic activity, but is also typically associated with thunderstorms and meteor showers. In the daytime these regions of sporadic E can unexpectedly fully or partially reflect oblique radio waves aimed at the F layer for frequencies up to 20 MHz [8]. This is one of the prominent causes of radio signal fading. In the high-latitude ionosphere it is commonly a nighttime phenomenon correlated to magnetic activity. In the polar cusp itself, sporadic E is generally quite weak and in fact has a negative correlation to magnetic activity [7].

F Layer [160 km - 400 km]: Electron density $10^{11} - 10^{12}$ m^{-3}

This is the layer with the highest electron density and is where most HF waves get refracted back towards the earth. It can be subdivided into a lower F1 layer and upper F2 layer. The F2 layer is the most important layer for reflecting HF radio waves because it has the highest electron density and exists throughout the day due to its slow discharge. It is unfortunately also the most variable ionospheric layer, with frequent anomalies resulting in scattering of reflected signals. This phenomena is known as Spread F and can be observed on the ionogram of the received signal, as a data spread in the reflection height or the F2 critical frequency supported by the layer [9].

2.1.2 Regular Ionospheric Variations

The properties of the ionosphere vary with season, solar cycle, geographical location and time of day. These variations can last anywhere from a few minutes to a few years. In addition to the regular solar and geomagnetic variations, ionospheric variations have been linked to atmospheric waves propagating from the lower to upper regions of the atmosphere. Examples of these are atmospheric gravity waves (AGWs), as well as tidal and planetary waves [10]. As already discussed in the introduction, the ionospheric sounding system considered in this **book** aims to detect and characterise the travelling iono-spheric disturbances caused by AGWs in particular.

Figure 2.1 shows the vertical electron density profiles at a midpoint between the South Pole and SANAE station generated using the International Reference Ionosphere (IRI) model (discussed further on) for different times and seasons during both solar maximum and solar minimum.

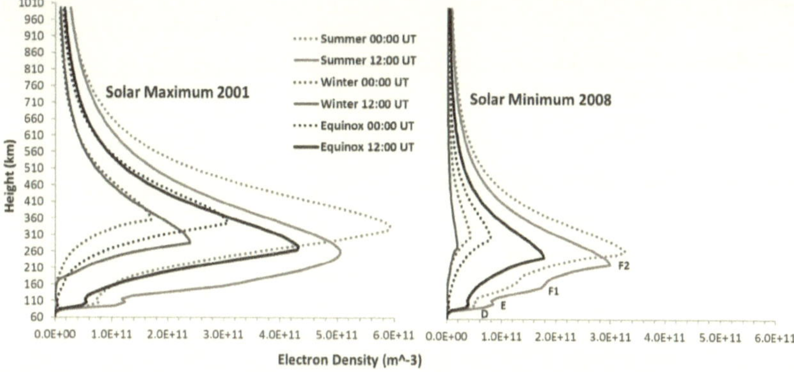

Figure 2.1: Electron density height profile for different seasons and time of day, for the solar maximum of 2001 (left) and the solar minimum of 2008 (right). Generated from IRI model at $80.833°S$ $2.7802°W$.

Seasonal and Diurnal Variations

The electron density in the ionosphere varies strongly with the position of the sun, consequently variations relating to the time of day and season occur. As a result of the lack of ionising solar radiation, the D and F1 layers almost completely disappear when the sun sets, whilst the E layer electron density becomes substantially weaker (See Appendix Figure C.1). In mid-latitudes, this results in stronger signals being received during the evening because less absorption of the radio wave occurs. In the polar regions however, this so-called day/night effect is warped because in summer the sun does not set, and in winter it does not rise. The day/night variation will, however, be present at times of equinox in the polar regions. Understandably, mid-summer electron densities will on average be higher than the other months. Interestingly from the sample profiles generated for the polar ionosphere in Figure 2.1, it appears that at midnight (00:00 UT) in summer, the electron densities in the F2 layer reach higher levels than at midday (12:00 UT). This is likely because although the solar zenith angle is nearly constant over a 24-hour period, there is still a daily variation in the electron density from high-latitude plasma convection which depends on the separation between the geographic and geomagnetic poles at the time [7].

Solar Cycle Variations

The sunspot number (SSN) and solar radio flux at 2800 MHz (F10.7) are the key indices used to measure solar activity [8]. The sun cycles in periods of roughly 11 years, where the solar maximum corresponds to high solar activity and a maximum number of sunspots observed on the sun. Similarly the solar minimum corresponds to low solar activity and a minimum number of spots observed on the sun. It can be expected that in summer, at solar maximum, the highest levels of regular ionisation will be observed. This is confirmed by the electron density profile of Figure 2.1.

2.1.3 Irregular Ionospheric Variations

Irregular ionospheric variations that occur are very hard to predict and can result in all communication being lost. Sporadic E and Spread F are two such examples which have already been mentioned. Ionospheric storms (geomagnetic, auroral or magnetospheric) are another well-known cause of disturbance. These have a strong link to solar emissions and usually result in a sudden large increase in electron density as well as changes in the height of the F2 layer.

2.1.4 The High-Latitude Ionosphere

The high-latitude ionosphere is highly vulnerable to the effects of spaceborne energetic particles due to the geomagnetic field lines connecting it to the outer magnetosphere. Figure 2.2 was extracted from [11] in order to demonstrate this. The high-latitude ionosphere area is made up of the auroral oval, F trough and polar cap. These areas all experience phenomena that can effect HF communication circuits extensively and are briefly described here.

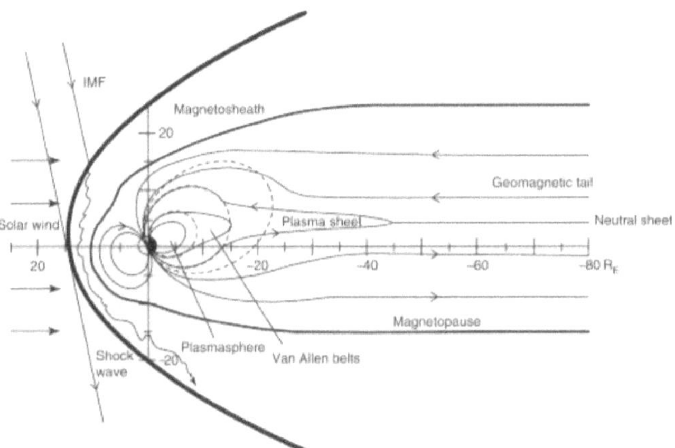

Figure 2.2: Illustration of the earth's magnetosphere for a southward Interplanetary Magnetic Field (IMF) extracted from [11].

The auroral oval consists of a ring centred on the geomagnetic poles which marks the boundary that separates open and closed magnetic field lines. Owing to the closed magnetic field lines, energetic particles are continuously entering the region, resulting in enhancements in electron density and ionisation levels. It is an area most commonly associated with natural light displays (aurora) accredited to particle precipitation in the E-region of the Ionosphere. Poleward of the auroral zones, dips in the ionisation levels known as high-latitude "troughs" occur in the F-layer. These F troughs have been correlated to a combination of plasma convection, precipitation and the effects of heating and neutral winds [12].

The polar cap is the region enclosed by the auroral oval that connects the geomagnetic field to the Interplanetary Magnetic Field (IMF). It is thus particularly vulnerable to the effects of incoming energetic particles, especially those associated with solar events. Some of the most prominent ionospheric features that can be observed in this region are patches and sun-aligned arcs that predominantly appear in the F-layer. Patches, which are transported to the polar caps via polar convection, are most frequently observed during winter and usually result in an increase in electron density levels. On the other hand, sun-aligned arcs are accompanied by a notable decrease in electron density and are correlated to periods of low geomagnetic activity.

2.1.5 Ionospheric Models

Combining the information supplied by ionospheric/atmospheric models with propagation formulations is essential to HF propagation planning. Models are usually categorised as theoretical, empirical or parametric. Theoretical models of the ionosphere require a lot of computational power in order to be able to solve highly complex equations. These equations cannot easily account for a disturbed ionosphere because some of the phenomena involved are too complicated and unpredictable to thoroughly quantify. This is particularly true in regions of increased geomagnetic activity.

Empirical models of the ionosphere are based on experimental data and have little dependency on theoretical predictions, therefore they are usually the best source of ionospheric parameters. However, empirical models require an extensive and reliable database, thus their accuracy is highly location-dependent. In remote areas, such as at high-latitudes, especially in the Southern Hemisphere, the accuracy of these models rapidly deteriorates due to a lack of observational data. Parametric models use simplified versions of the theoretical equations combined with experimental data. For this reason, they are also referred to as semi-empirical models. However, these models are still confined to well-understood geophysical phenomena and frequently observed areas.

The International Reference Ionosphere (IRI) model is a semi-empirical model developed and annually upgraded by the Committee on Space Research (COSPAR) and the International Union of Radio Science (URSI). The IRI obtains the majority of its ionospheric data from a global network of ionosondes and incoherent scatter radars. Additionally, data is obtained from topside measurements using satellites and rockets. For areas which do not have a lot of coverage, theoretical and parametric techniques are exploited to fill the gaps. For a given time, date and location, the IRI provides a monthly average of the various ionospheric parameters for a given altitude in 1 km minimum range intervals.

Good accuracy has been reported for the IRI model at mid- and low-latitudes. However, its reliability has been disputed at high-latitudes where there is station scarcity and an exceptionally variable ionosphere [13]. The IRI model should thus be used cautiously for application in Antarctica, especially when a disturbed ionosphere is considered. That being said, the use of the IRI model was still sufficient for the purpose of an overall propagation path analysis in quiet ionospheric conditions and was used later on in this report.

Other Applicable Models:

The other prominent semi-empirical models recommended by the Community Coordinated Modeling Center (CCMC) for international use are:

1. The International Geomagnetic Reference Field (IGRF) model developed by the International Association of Geomagnetism and Aeronomy (IAGA) [14]. Year-long magnetic field flux densities and magnetic inclination angles can be obtained from this model.

2. The global reference atmospheric model (NRLMSISE-00) developed by the US Naval Research Laboratory (NRL) [15]. The neutral density of individual atmospheric components can be obtained from this model.

2.2 Magnetoionic Theory

Magnetoionic theory details how an electromagnetic wave will behave when propagating through an ionised magnetic medium, such as the ionosphere. It is essential for determining the complex refractive index of the wave,

$$n = n_r - j\chi \qquad (2.2)$$

needed for ray tracing and absorption predictions. The main effect of the geomagnetic field is that it splits an incident radio wave into two waves with slightly different refractive indices [9]. Most ray tracing models simply neglect the geomagnetic field because at high operating frequencies its effect becomes minimal. However, especially at lower operating frequencies, by neglecting the magnetic field you may limit the accuracy of propagation paths and losses predicted by models. In a system with transmitter power constraints, any unforeseen absorption not included in a typical fade margin may result in the received signal scarcely missing the detection threshold.

This section discusses the prominent formulations used for the refractive index of a wave in a magnetoplasma, and describes the simplifications frequently made in ray tracing. It then details the integrals used to determine non-deviative absorption in the lower ionosphere. This absorption is the most prominent propagation loss after that of free-space.

2.2.1 The Appleton-Hartree Dispersion Relations

The formulations for the refractive indices most commonly used in models are those of Appleton-Hartree and Sen-Wyller. Appleton-Hartree is the classical formulation and is still the most frequently utilised in ray tracers because of its simplicity and fast computational implementation. The more recent Sen-Wyller formulation is a generalisation of the Appleton-Hartree equations with a key difference being that it does not consider the electron collision frequency independent of electron energy. In general, literature considers the Sen-Wyller equations to be the more accurate approximation in the non-deviative absorbing D and E ionospheric layers, while the Appleton-Hartree equations are considered to be more accurate in the F layer. It has previously been assumed that the Appleton-Hartree equations should not be used in the D/E layers. However, [16] has determined that if the Schunk-Nagy definition of collision frequency (given later) is utilised, the difference in absorption calculated becomes less than 5%.

As a result, only the original Appleton-Hartree formulation for the complex refractive index of a wave with angular frequency ($\omega = 2\pi f$) was considered and is given by,

$$n^2 = 1 - \frac{2X(1 - jZ - X)}{2(1 - jZ)(1 - jZ - X) - Y_T{}^2 \pm \sqrt{Y_T{}^4 + 4Y_L{}^2(1 - iZ - X)^2}} \tag{2.3}$$

$$X = \frac{\omega_p^2}{\omega^2} \tag{2.4}$$

$$Y = \frac{\omega_H}{\omega}, \quad Y_T = Y \sin\theta_H, \quad Y_L = Y \cos\theta_H \tag{2.5}$$

$$Z = \frac{v_e}{\omega} \tag{2.6}$$

$$\omega_p^2 = \frac{N_e q_e^2}{m_e \epsilon_0} \quad and \quad \omega_H = \frac{B q_e}{m_e} \tag{2.7}$$

where N_e $[m^{-3}]$ is the electron density, B [T] is the flux density of the magnetic field, v_e $[s^{-1}]$ is the collision frequency, and θ_H is the angle between the propagating wave and the geomagnetic field. The rest are fundamental constants as defined in the list of units and symbols. The plasma frequency (ω_p) represents the angular frequency of an electron's simple harmonic motion in a plasma (cold weakly ionised gas) [9]. The angular gyrofrequency (ω_H) represents the angular speed of a charged particle in the presence of a magnetic field.

The electron collision frequency (v_e) in equation (2.6) is used to quantify collisional absorption and is a combination of both electron-neutral (v_{en}) and electron-ion (v_{ei}) collisions. For ease of the calculations done in this book, electron-ion collisions are neglected because they are negligible in comparison to electron-neutral collisions in regions below 120 km. The Schunk-Nagy electron-neutral collision frequency suitable for use in the Appleton-Hartree equations is defined as a weighted function of the different contributions from neutral species in the ionosphere [17]:

$$v_{en} = v_{eN_2} + v_{eO_2} + v_{eO} + v_{eHe} + v_{eH} \tag{2.8}$$

with,

$$v_{eN_2} = (2.33 \times 10^{-11})n[N_2](1 - (1.21 \times 10^{-4}T_e))T_e \tag{2.9}$$

$$v_{eO_2} = (1.82 \times 10^{-10})n[O_2](1 + (3.6 \times 10^{-2}T_e{}^{0.5}))T_e{}^{0.5} \tag{2.10}$$

$$v_{eO} = (8.9 \times 10^{-11})n[O](1 + (5.7 \times 10^{-4}T_e))T_e{}^{0.5} \tag{2.11}$$

$$v_{eHe} = (4.6 \times 10^{-10})n[He]T_e{}^{0.5} \tag{2.12}$$

$$v_{eH} = (4.5 \times 10^{-9})n[H](1 - (1.35 \times 10^{-4}T_e))T_e{}^{0.5} \tag{2.13}$$

In these equations, the electron temperature (T_e) is in kelvins and the density of neutral particles ($n[\ ]$) in cm^3. In the higher regions of the ionosphere, where refraction begins to take place, collisions decrease significantly. For this reason, the collision frequency in the F layer can be neglected, and the non-collisional approximation (i.e. $Z = 0$) of the Appleton-Hartree equations can be used. The refractive index is then purely real (i.e. $n = n_r$). In the D and E layers, the electron collision frequency is higher as a result of the dense surrounding neutral atmosphere and should not be ignored. This is why radio wave absorption mostly occurs in the D and lower E layers and not in the upper ionospheric layers.

In general, the effects of the geomagnetic field can be neglected for operating frequencies much greater than the gyrofrequency. In the South Pole, the gyrofrequency usually ranges between 1.5 MHz to 2 MHz (obtained using the IGRF model), which is significant enough relative to 12.57 MHz to not be ignored. The presence of the geomagnetic field splits the incident wave into an ordinary (O) wave mode and extraordinary (X) wave mode, denoted respectively by the plus or minus sign in the denominator of equation (2.3). This results in two separate waves with different refractive indices. The relative power delivered to each mode is related to the angle of the propagating ray with respect to the geomagnetic field (θ_H). Knowledge of the relative power of the two propagation modes (X and O) can greatly assist in the interpretation of received signals. Consider the following scenarios for the non-collisional approximation:

1. If a ray travels parallel to the magnetic field (i.e. $\theta_H = 0$) it is referred to as quasi-longitudinal (QL) propagation and yields,

$$n_r{}^2 = 1 - \frac{X}{1 \pm Y} \tag{2.14}$$

   An example of this type of propagation would be a vertically incident wave at the magnetic poles. According to derivations in [9], if a linearly polarised antenna is used in this example, equal power would then be coupled to both the ordinary and extraordinary mode.

2. If a ray travels perpendicularly to the magnetic field (i.e. $\theta_H = 90°$), it is referred to as quasi-transverse (QT) propagation and yields,

$$n_r{}^2|^O = 1 - X \quad and \quad n_r{}^2|^X = 1 - \frac{X(1-X)}{1 - X - Y^2} \tag{2.15}$$

   for the ordinary and extraordinary modes respectively. It should be noted that the refractive index of the ordinary mode in this case is the same as it would be if you neglected the magnetic field altogether. An example of this type of propagation would be a vertically incident wave at the magnetic equator. If a linearly polarised antenna orientated in the north-south direction is used in this case, all power would be coupled to the ordinary mode, according to derivations detailed in [9].

Since the system dealt with in this book utilises an obliquely incident wave propagating over a large distance, a more complex analysis is necessary in order to make any comprehensive assumptions about the amount of power that would likely be received from each mode. This derivation is beyond the scope of this report. However, similar research has been done for an experiment using a high-latitude SuperDARN at 5 MHz (Saskatoon) in the Northern Hemisphere, where it was found that the extraordinary mode dominates for a large band of azimuths [18]. This experiment had a different geometry and used a lower operating frequency (thus more vulnerable to the effects of the geomagnetic field) than the intended South Pole to SANAE system, and is mentioned purely to highlight that the extraordinary mode can be significant at high-latitudes.

2.2.2 Simplified Ray Theory

The simplest form of ray tracing takes place when the effects of both the magnetic field and collisions are ignored in equation (2.3) (i.e. Y=0 and Z=0). An abridged version of the refractive index of a wave at frequency (f) is then given by,

$$n_r{}^2 = 1 - X = 1 - \frac{N_e q_e^2}{m_e \epsilon_0 (4\pi^2 f)} \approx 1 - \frac{81 N_e}{f} \tag{2.16}$$

As mentioned, this is equivalent to QT propagation of the ordinary wave. If we consider the ionosphere as layers of uniform electron density increasing with altitude, it can be seen from equation (2.16) that the refractive index will decrease as the wave travels through it. Applying Snell's law of refraction, this means that the wave will slowly be bent back down towards the earth until it reaches a point of total reflection. For a ray to reach this point of return, there must be a layer which results in a refractive index equal to the sine of the wave's angle of incidence relative to the vertical ($\phi_{i,n}$) i.e.

$$sin(\phi_{i,n}) = n_r = \sqrt{1 - \frac{81 N_e}{f}} \tag{2.17}$$

The frequency at which the condition of (2.17) is met for the ordinary mode vertically incident wave ($\phi_{n,i} = 0$) is known as the critical frequency and is given by,

$$f_O = 9\sqrt{N_{max}} \tag{2.18}$$

The F2 layer usually has the highest electron density (N_{max}) and is thus the layer at which most HF waves are reflected. Therefore, the F2 ordinary wave mode critical frequency ($foF2$) is most commonly quoted. The following secant law can be used to find the equivalent oblique frequency (f_{ob}) of a wave entering a flat ionosphere at an angle of incidence (ϕ_i), needed to reach the same height of ionospheric reflection as a vertically incident wave with frequency (f_v) [9]:

$$f_{ob} = k f_v \sec \phi_i \tag{2.19}$$

where k is a correction factor typically added to account for the curvature of the earth. For ground distances greater than 500 km a k value of 1.115 is recommended [9]. These equivalent frequency relations are demonstrated by Figure 2.3.

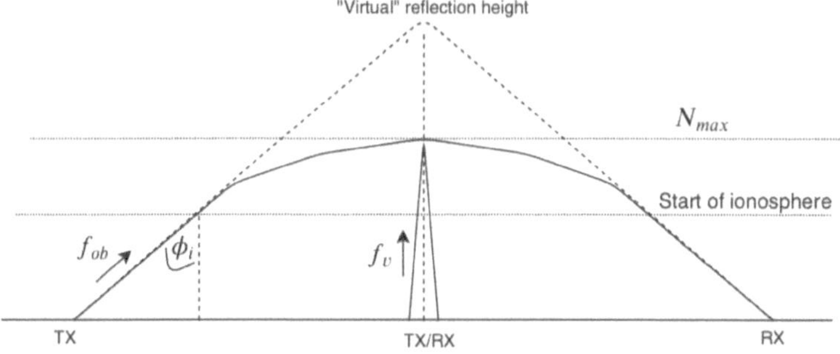

Figure 2.3: Illustration of the equivalent vertical and oblique waves reflected from the ionosphere at the same height.

Now, if the curvature of the earth is taken into account, as well as a thin ionospheric layer assumed [9], the maximum usable frequency (MUF) supported for skywave propagation can be estimated using the geometry in Figure 2.4. This is done using the propagation factor ($\sec \phi_0$) in conjunction with the F2 critical frequency via:

$$MUF = f_oF2 \times \sec \phi_0 \tag{2.20}$$

This can be translated to a function of the earth's radius (R), propagation ground range distance (D) and reflection height (h_0):

$$MUF = f_oF2 \times \sqrt{\frac{\frac{D^2}{4} + \left(h_0 + \frac{D^2}{8R}\right)^2}{\left(h_0 + \frac{D^2}{8R}\right)^2}} \tag{2.21}$$

Often the MUF and propagation factor are quoted for a ground range distance of 3000 km which are called $MUF(3000)F2$ and $M(3000)F2$, respectively. This corresponds to a lowest ground take-off angle of 3° and is the practical limit for single-hop propagation. Once again, let it be emphasised that the simple ray tracing model discussed above only takes the ordinary mode critical frequency $foF2$ into account. As a result, the MUF can often be predicted to be lower than what it actually is [19]. For propagation planning, this could impose unnecessary limitations.

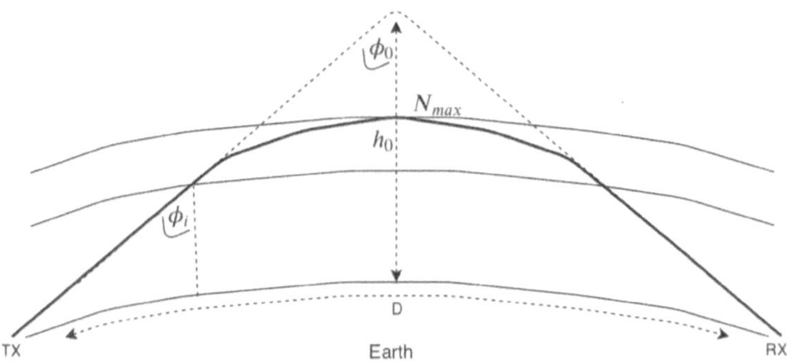

Figure 2.4: Simplified ray tracing assumed for a curved earth and a thin ionospheric layer.

2.3 Ionospheric Absorption of HF Radio Waves

A propagating radio wave will experience a loss in power along its path caused by multiple factors, the greatest being free-space attenuation. Another prominent loss is the ionospheric absorption that occurs when electrons collide with neutral molecules and ionised particles, resulting in some of the wave's energy being dissipated into heat and electromagnetic noise [9]. This happens because electrons in the ionosphere respond to radio waves' oscillating electromagnetic fields, while heavier ions remain stationary. When ionospheric conditions are considered normal and spatial spreading is neglected, the total ionospheric absorption loss in decibels can be described by,

$$L_a = -8.68 \int k.ds \tag{2.22}$$

where k is the absorption coefficient obtained from the imaginary part χ of the complex refractive index using,

$$k = \frac{\omega \chi}{c} \tag{2.23}$$

Using the Appleton-Hartree dispersion relations and ignoring the presence of the geomagnetic field, the absorption coefficient is approximated in dB/km by,

$$k = \frac{8.68 q_e^2}{2 m_e c \epsilon_0 n_r} \frac{N_e v_e}{\omega^2 + v_e^2} = \frac{4.6 \cdot 10^{-2}}{n_r} \frac{N_e v_e}{\omega^2 + v_e^{\,2}} \qquad \text{[dB/km]} \tag{2.24}$$

Recall that, n_r is the real part of the complex refractive index, $N_e \; [m^{-3}]$ is the electron density, and $v_e \; [s^{-1}]$ is the electron collision frequency. The rest are fundamental constants as defined in the list of units and symbols. The above equation demonstrates that higher frequency waves are absorbed less.

2.3.1 Non-Deviative D/E Layer Absorption

Absorption can be divided into two types, namely non-deviative and deviative absorption. Non-deviative absorption occurs when there is little to no refraction of the section of ray (i.e $n_r = 1$), which can be assumed in the D and lower E layers where the electron density is fairly low. Deviative absorption occurs higher up in the ionosphere (usually in the F layer) or wherever significant bending of the ray takes place. Non-deviative absorption is the much greater of the two and is hence the only type considered in this report. In the presence of the earth's magnetic field, the non-deviative absorption coefficient is [9],

$$k = 4.6 \times 10^{-2} \frac{N_e v_e}{(\omega \pm \omega_H \cos \theta_H)^2 + v_e^{\,2}} \qquad \text{[dB/km]} \tag{2.25}$$

Once again, θ_H is the angle between the propagating wave and geomagnetic field and ω_H is the angular gyrofrequency. The term $\omega_H \cos \theta_H$ is effectively the component of the angular gyrofrequency in line with the geomagnetic field and is added for ordinary waves (O mode) and subtracted for extraordinary waves (X mode). The above equation indicates that the absorption of the X mode is greater than that of the O mode. This difference increases as the gyrofrequency approaches the wave frequency and the X mode is almost completely absorbed [9].

The non-deviative absorption loss of a vertically incident wave on the ionosphere can be found using the following integral with 1 km height increments (i.e. dh=1 km),

$$L_a^v(\omega_v) = 4.6 \cdot 10^{-2} \int_{h_1}^{h_2} \frac{N_e v_e}{(\omega \pm \omega_H \cos \theta_H)^2 + v_e^{\,2}} .dh \qquad \text{[dB]} \tag{2.26}$$

The angular frequency of a wave vertically incident on the plane ionosphere (ω_v) can be related to the angular frequency of an equivalent oblique wave (ω_{ob}) at the angle of incidence (ϕ_i) to the vertical by,

$$\omega_v = \omega_{ob} \cos \phi_i \tag{2.27}$$

Subsequently, Martyn's theorem says that the absorption of the oblique wave (L_a^{ob}) can be related to the absorption of the equivalent vertical wave (L_a^v) with angular frequency ω_v by [20],

$$L_a^{ob}\big|_{\omega_{ob}=\omega} = L_a^v\big|_{\omega_v=\omega_{ob}\cos\phi_i} \cos \phi_i \tag{2.28}$$

The oblique wave's increase in absorption from travelling through the ionosphere for longer is compensated for by a decrease in absorption due to the oblique wave having a higher frequency than the equivalent vertical wave. The oblique absorption can now be calculated using the following integral,

$$L_a^{ob}(\omega) = \frac{4.6 \cdot 10^{-2}}{\sec \phi_i} \int_{h_1}^{h_2} \frac{N_e v_e}{(\omega \cos \phi_i \pm \omega_H \cos \theta_H)^2 + v_e^2} . dh \qquad [\text{dB}] \qquad (2.29)$$

in which θ_H corresponds to the angle between the geomagnetic field and the equivalent vertical wave, as opposed to the angle between the oblique wave and the magnetic field. A simplified diagram of all the relevant parameter definitions and assumptions is shown further on in Figure 4.6 in the chapter where these equations are implemented.

2.4 High-Latitude Absorption Events

D layer absorption in steady ionospheric conditions will reduce signal strength to an extent, but will rarely do so enough to prevent communications, except in the event of ionospheric disturbances. In the polar regions, there are absorption phenomena which occur sporadically and result in significant augmented D layer absorption. The two most important of these, namely polar cap absorption (PCA) and auroral radio absorption (AA), are discussed in this section.

To provide some insight into the nature of these absorption events, previous studies were investigated and applied to this project to get an idea of the potential impact on the system's functioning. It should be noted that the absorption studies which were considered used a 30 MHz vertically incident riometer to measure absorption levels. Riometers generally use the band 30 MHz to 50 MHz, as lower frequencies are heavily absorbed and require larger antennas which are more susceptible to interference. The riometer indicates an absorption event by comparing measured levels of absorption against a quiet-day curve (QDC), which is a seasonal-dependent prediction of the daily signal level in the absence of absorption [7]. The starting time of an absorption event is taken as being when the level of the 30 MHz wave ionospheric absorption exceeds the QDC by more than 0.5 dB [21]. In order to interpret 30 MHz riometer measurements of absorption relative to a 12.57 MHz wave, they need to be frequency-scaled according to the inverse square law (in most circumstances),

$$L_a(f) \approx L_a(30MHz) \times (\frac{30}{f})^2 \qquad (2.30)$$

It should be noted that the above equation should technically use an effective frequency related to the mode (X or O) of propagation by adding or subtracting the gyrofrequency. Furthermore, although this **book** concerns oblique propagation, only vertical inci-dence absorption is analysed here. The calculated scaled absorption is thus not necessarily an accurate estimation of the absorption that could be seen in a 12.57 MHz oblique sound-ing system. However, it will suffice for the purpose of giving some general indication of the impacts of these high-latitude events.

2.4.1 Polar Cap Absorption Events

Emissions of energetic solar protons (usually associated with solar flares) have rapid and easy access to the polar atmosphere through the near vertical geomagnetic field connecting it to interplanetary medium (material which fills the Solar System). These emissions result in additional ionisation in the ionosphere. Hence there is an increase in HF radio wave absorption for a period of time which is known as a polar cap absorption event [21]. The duration of these PCA events can last anywhere between an hour to several days, with their occurrence depending strongly on the stage of the solar cycle. For instance, there are sometimes more than ten per year near solar maximum and sometimes none (or very few) near solar minimum. The long term yearly average is approximately six events [7]. Concerning PCA specifically, the frequency-dependence relation in (2.30) has been determined to be with an exponent of 1.5 instead of 2 and thus,

$$L_a(f) \approx L_a(30MHz) \times (\frac{30}{f})^{1.5} \tag{2.31}$$

This exponent was found through analysis of multi-frequency absorption data taken during solar proton events (SPE) [22] and is in good agreement with calculated theoretical daytime exponents in the range 1.4 to 1.6 done by [23]. This applies for 24-hours in polar summer months, given that the sun does not set.

Three different PCA events of different magnitude and period were recorded over the solar maximum in November 2001 at Terra Nova Bay in Antarctica, the results of which can be seen in Figure 2.5 extracted from the published findings [21]. The first and third events were similar in nature and extended over about five days, reaching peak magnitudes of 18 dB at 30 MHz which corresponds to ≈ 66 dB at 12.57 MHz using (2.31). The second and smallest event only lasted one day with a maximum of ≈ 4 dB at 12.57 MHz equivalent to 1 dB at 30 MHz. These findings confirm that a medium to high PCA event over the path between SANAE and the South Pole will probably result in communication being lost. This is because a low-powered system is to be implemented and is unlikely to be able to compensate for large unforeseen absorption.

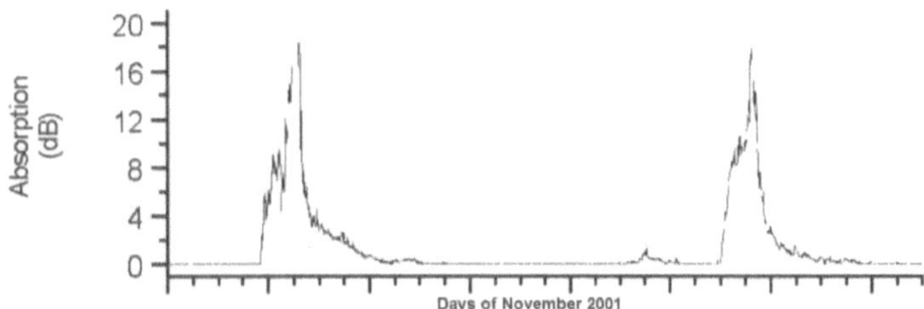

Figure 2.5: Polar cap absorption (PCA) events recorded with a 30 MHz riometer at Terra Nova Bay in Antarctica during November 2001 extracted from [21].

2.4.2 Auroral Radio Absorption

As the name suggests, this type of absorption only occurs in auroral zones and is a result of sporadic energetic electron precipitations from the magnetosphere during auroral episodes [7]. The path between the South Pole and SANAE station crosses an Auroral zone, so it is likely that the path availability may at times be effected by the resulting radio wave absorption. Figure 2.6 shows the auroral absorption recorded by a vertically incident 30 MHz riometer on 15 October 1963 at Byrd station, Antarctica and was extracted from

[24]. It can be seen from these recordings, that the absorption characteristic of an AA event tends to occur in sudden bursts and will ordinarily only last from tens of minutes to an hour [7]. Although, these spikes can continue for longer. The magnitude of absorption seen is not usually as severe as that of PCA events. These AA events will normally occur several times a day, causing intermittent interruptions in the level of received signal.

In the specific set of data shown in Figure 2.6, the peak absorption recorded was about 9 dB at 30 MHz and subsided after an hour. This corresponds to $\approx 51\ dB$ at 12.57 MHz using the relation in equation (2.30). Owing to their smaller magnitude and period, AA events are unlikely to always be detrimental to HF communications. Therefore even when large peaks occur, there will unlikely be continuous disruption for an extended period.

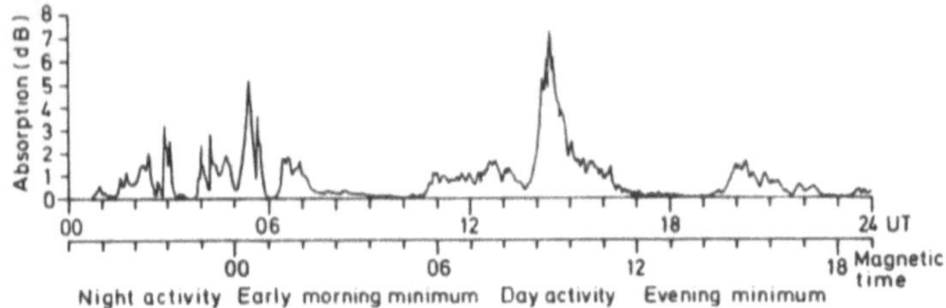

Figure 2.6: Typical daily Auroral Absorption (AA) events recorded with a 30 MHz riometer at Byrd station in Antarctica on 15 October 1963 [24]

2.5 HF Propagation Prediction Models

Ray tracing is complicated by the fact that the magnitude of ionospheric properties vary with the direction of measurement (anisotropic medium). Accurate methods of modelling require step-by-step integration of various simultaneous equations and have been discussed in [25]. One such commonly used advanced ray tracing algorithm is the Jones-Stephenson model [26]. Another highly esteemed interface is the Advanced Refractive Effects Prediction System (AREPS) developed by the United States Navy, which extracts data directly from ionospheric and geophysical models for use in computations. AREPS includes the effects of the geomagnetic field and has very useful c apabilities. However, it is not easy for general users to gain licensed access to it because of its US military links. For the purpose of being able to easily follow up on any studies done in this book, only freely available ray tracers that use the simplified ray tracing model previously discussed were considered.

The primary models reviewed for use in propagation predictions were the Voice of America Coverage Analysis Program (VOACAP), Ionospheric Communication Enhanced Profile Analysis and Circuit program (ICEPAC), and Recommendation 533 (REC533). All of these models were developed by the Institute for Telecommunications (ITS) but REC533 is the model recommended and updated by the International Telecommunication Union (ITU). Although global predictions can be made with these programs, unfortunately, because models specific d eveloped f or t he S outhern H emisphere a re s everely l acking, any predictions made are likely to be unreliable. Furthermore, the capabilities of these programs are generally limited to quiet ionospheric conditions, with prediction accuracy falling to about 45% in disturbed conditions [27].

The models use worldwide maps of monthly median values for the critical frequency $f0F2$ and propagation factor $M(3000)F2$. In VOACAP and ICEPAC these ionospheric coefficients can be set to those provided by either the Consulate Committee on International Radio (CCIR) or the International Union of Radio Science (URSI). The recommended default parameters are the CCIR coefficients, which are based on data recorded between 1954-1958 by a global network of 150 ionosondes and are also those used by the IRI model [28]. REC533 uses a quick-run ionospheric model called NeQUICK ionospheric model.

VOACAP was completed in 1993 and is an improvement of the original IONCAP program first developed in 1983. VOACAP was considered inadequate for predictions in areas with a lot of geomagnetic activity, specifically in the high-latitude ionosphere and was consequently modified to ICEPAC. ICEPAC includes the Ionospheric Conductivity and Electron Density (ICED) model which is a statistical model of the important individual characteristics of the ionosphere at different locations. It is based on observations of the Northern Hemisphere ionosphere and contains specific algorithms for various geographical regions of the high-latitude ionosphere, such as the polar cap and auroral trough to name a few [29]. ICEPAC also allows for a magnetic activity index Q_e to be put in to simulate disturbed ionospheric conditions. However, this index should be used with caution and is in fact not recommended for inexperienced users [30]. Owing to the high-latitude capabilities of ICEPAC, it was decided that it would be the most suitable program for predictions in the South Pole ionosphere. ICEPAC can be used for predicting transmission loss, MUF and optimum take-off elevation angle, among others.

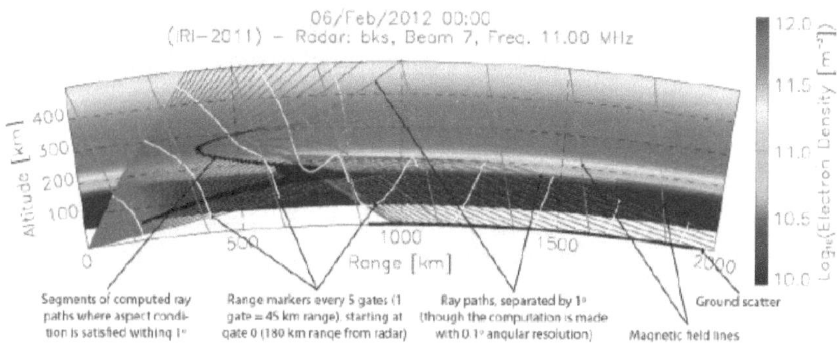

Figure 2.7: Example of the VT online ray tracing tool output extracted from [31].

In addition to the aforementioned ray tracers, Virginia Tech (VT) University developed a ray tracer specifically for use with the SuperDARN radars. This online tool computes the ordinary wave ray paths by solving the simple 2-dimensional ray tracing scheme from [32] using a Runge-Kutta numerical technique and ionospheric profiles generated using the IRI-2011. The simplified Appleton-Hartree refractive index referred to in equation (2.16) is used [31]. An example of the output of this ray tracer is shown in Figure 2.7. Graphs of elevation angle brackets and relative power as a function of range can be determined, however only graphical data and not specific data is available using this tool.

2.6 Chapter Summary

The intention of this chapter was to give a brief overview of HF ionospheric propagation and the parameters that should be examined when propagation planning. It also provided some insight into the high-latitude ionosphere and its particular properties. An in depth analysis of the formulations for the refractive index was done in order to build towards being able to better quantify the absorption that a propagating radio wave may see over its path to the receiver at SANAE. The classic Appleton-Hartree dispersion relations were selected for use. In quiet ionospheric conditions, it was established that non-deviative absorption in the D and lower E layers will be the largest propagation loss after that of free-space loss. The integrals for estimating this absorption for an oblique wave were subsequently developed using various sources. Relevant high-latitude D and E layer absorption events were also researched in order to get an idea of their potential impact on the 12.57 MHz radio wave under consideration. This was accomplished by briefly analysing recordings (obtained from literature) of specific polar cap absorption (PCA) and auroral radio absorption (AA) events that occurred in Antarctica.

This chapter highlighted some of the shortcomings of available ionospheric models and ray propagation prediction tools. One of the main findings was that, in general, ray tracers ignore the effects of the geomagnetic field and assume the propagation and absorption of the ordinary wave only. The International Telecommunication Union Recommendations (ITU-R) [33] conveys that, at high-latitudes, the extraordinary wave mode will have horizontal polarisation at highly oblique angles of incidence on the ionosphere. This finding is relevant because the $\approx$ 2090 km ground range of the single-hop skywave link will require fairly low take-off angles. Furthermore, the SuperDARN receiver antenna is only horizontally polarised and will therefore be well-matched to the incoming extraordinary modes.

Taking this into consideration, along with the fact that the extraordinary modes are strongly absorbed, there is motivation to ensure that any link analysis done accounts for the extraordinary mode ionospheric loss. Doing this will help to ensure that enough transmitter power and antenna gain is specified such that both the ordinary and extraordinary modes arrive with enough power to be detected by the receiver. This would assist in increasing the reliability of the HF skywave system being designed.

3 HF Antenna Considerations

An antenna is a device which converts bound circuit fields into propagating electromagnetic waves and, in reverse, can receive power from passing electromagnetic waves. For the point-to-point skywave link being dealt with here, the chief design objective was to produce as little radiation behind the antenna as possible so as not to interfere with an already existing South Pole SuperDARN array. This would require a directional antenna, which radiates/receives more power in certain directions. The result of this is an improvement in the specified transmission performance and a reduction in interference.

This chapter starts off by exploring a few of the basic antenna parameters that need to be considered in the design process. Antenna polarisation was rigorously discussed and research into the specific South Pole ground plane medium was undergone. Following which, a directional HF antenna literature review was done to narrow down a list of appropriate prospective antennas. The selected antennas were analysed in depth for rudimentary implementation in the design chapter.

3.1 Basic Antenna Parameters

3.1.1 Antenna Efficiency, Directivity and Gain

The efficiency (η) of an antenna can be defined as the ratio of effectively radiated power (P) to the total input power of the antenna [34]. It only includes the power loss due to the resistance of the antenna, namely P_{res} (i.e. it does not include losses caused by impedance and polarisation mismatches). Accordingly,

$$\eta = \frac{P}{P + P_{res}} \tag{3.1}$$

The directivity (D) of an antenna is a dimensionless value which is usually defined as the ratio of the radiation intensity in a particular direction to the radiation intensity averaged over all directions relative to a sphere. Gain (G), on the other hand, is defined as the ratio of the radiation intensity in a particular direction to the radiation intensity that would be realised by an ideal isotropic antenna. Gain is essentially a combination of the antenna's directivity and antenna efficiency factor (η) and is usually expressed in decibels relative to an isotrope (dBi). It is given by,

$$G_{[dB]} = G_{[dBi]} = 10 \log \eta D \tag{3.2}$$

Although gain can sometimes be expressed relative to the gain of a lossless half-wave dipole (dBd), in this book only dBi is referred to.

3.1.2 Take-off Angle and Beamwidth

The angle of elevation, relative to the ground, at which maximum radiation occurs is known as the take-off angle and is denoted here by $\theta_{el,max}$. We are dealing with a skywave circuit on the brink of medium to long range (≈ 2090 km). For this range, elevation angles ranging from $10° - 45°$ are typically needed [35]. In a directional radiation pattern, the beamwidth is defined as t he a ngle b etween t he h alf-power p oints o f t he m ain lobe. In decibels, this corresponds to points on the main lobe which are 3 dB less than the maximum gain. The main lobe beamwidth can be characterised by two polar co-ordinate parameters; half-power elevation beamwidth ($\Delta\theta_{el}$) and half-power azimuth beamwidth ($\Delta\theta_{az}$). Both of these are centred on the take-off angle $\theta_{el,max}$ (Figure 3.1).

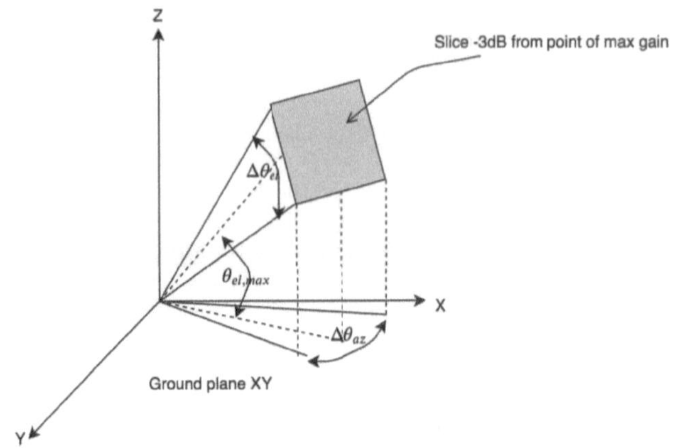

Figure 3.1: Antenna radiation beamwidth parameters in polar coordinates.

3.1.3 Front-to-Back Ratio

In a directional antenna, the front-to-back (FTB) ratio is defined as the ratio of gain in the maximum forward direction to the gain in the exact opposite direction (Figure 3.2). Once again, this parameter is usually quoted in decibels. Owing to the fact that often the antenna back lobes are not 180° to the main front lobe, it is important to consider what the radiation gain is for a range of angles behind the antenna and not only the FTB ratio of standard definition. This is especially so for HF horizontally polarised antennas because radiation reflected off of the earth below results in the antenna lobes being formed at an angle to the ground.

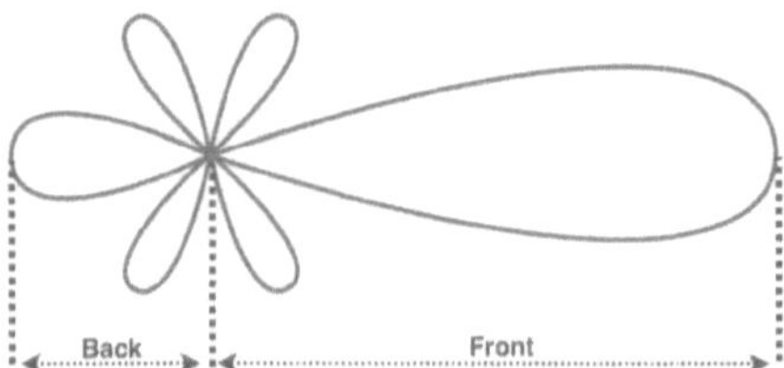

Figure 3.2: Simple diagram of the front-to-back ratio in a directional radiation pattern.

An alternate definition is given by [36] where the FTB ratio is defined as the ratio of the maximum direction gain to the peak gain within ±30° relative to a 180° angle from the maximum direction. In this particular project, there is more interest in the maximum gain in the entire region behind the antenna. Therefore, the FTB ratio itself is not necessarily an accurate reflection of the amount of radiation potentially interfering with the station at the South Pole. A clear definition of the antenna back lobe radiation specifications are given later in the antenna design chapters.

3.1.4 Bandwidth

Antennas can be roughly categorised into two types: Broadband (non-resonant) and narrowband (resonant) antennas. Narrowband antennas can usually only efficiently radiate energy within 2% of their resonant frequency while broadband antennas can still function over a wide range of frequencies (but less efficiently) [37]. Single frequency operation has already been specified for the system. Hence mainly resonant antennas were taken into consideration for the design. All antennas were optimised for radiation at 12.57 MHz frequency.

3.1.5 Antenna Feed and Impedance Matching

An antenna needs to be effectively impedance matched to the transmission line feeding it in order to minimise reflections and function with maximum efficiency at the required operating frequency. The measure of how well an antenna is impedance matched is given by the voltage standing wave ratio (VSWR) which is a function of the voltage reflection co-efficient (Γ) given by,

$$\Gamma = \frac{Z_L - Z_0}{Z_L + Z_0} \tag{3.3}$$

in which Z_L is the characteristic impedance of the antenna and Z_0 is the characteristic impedance of the transmission line. The VSWR can then be calculated using,

$$VSWR = \frac{1 + |\Gamma|}{1 - |\Gamma|} \tag{3.4}$$

A VSWR greater than 1 implies that there is an imperfect match and some power is being reflected back towards the source. However, a VSWR up to 2.5 can be tolerated provided that enough power is supplied. There are various impedance matching techniques that can be applied in order to improve the VSWR. Examples of such are "L" and "split capacitor" circuit networks, balun transformers, matching stubs and quarter wavelength transformers [38].

Antenna feed lines commonly utilise an unbalanced 50Ω coaxial cable because they are unaffected by nearby conductors and changes in environment. Coaxial cables are also very low in loss, even over long distances. The problem is that many antennas require a balanced feed to operate successfully. This results in a mismatch between the balanced and unbalanced elements causing a secondary current to flow on the outside of the coax. The simplest solution to this is to use an antenna balun. A balun is a transformer that converts an unbalanced signal (just a single conductor and a ground line) to a balanced one (two conductors with equal currents in opposite directions). In addition, baluns can also be designed to provide an impedance transformation which alleviates the need to provide a separate matching circuit. There are many baluns readily available on the market and so the specific design of one was not necessary here. It should be noted that impedance transformer baluns are usually specified in ratios with the load specified first and the input second. For example, an antenna with a 100 Ω characteristic impedance requiring a balanced feed from a 50 Ω coaxial would need a 2:1 antenna balun.

3.2 Antenna Polarisation for HF Skywave Applications

In order to transfer maximum power from a transmitter antenna to a receiver antenna, the characteristic wave polarisation of the antennas should match. However, for a linearly polarised wave incident on the ionosphere, the presence of the geomagnetic field usually causes the wave to become elliptically polarised. As a result, the polarisation of the received signal fluctuates and will differ from the polarisation of the transmitted signal. Hence, matching the antenna's polarisation does not necessarily improve the amount of signal being received [39].

The receiver SuperDARN antenna is purely horizontally polarised. According to [40], this likely results in increased fading seen on the received signal and investigations are being made to introduce SuperDARN antennas with dual polarity. In most cases, it is tricky to predict what the dominant polarisation mode will be upon reception of a signal. This arrival polarisation will mostly depend on the propagating signal's orientation relative to the geomagnetic field. At high-latitudes, where the geomagnetic field is nearly perpendicular to the ground, [9] determined that an emerging ordinary wave would dominantly be vertically polarised and hence a horizontally polarised receiving antenna would not be the most efficient choice. Nonetheless, the SuperDARN network has never faced any serious performance issues resulting from this.

3.2.1 Vertical vs. Horizontal Polarisation

The appeal of implementing a vertically polarised antenna is in the relative reduction of element size for the 24 m operational wavelength required. For example, a typical horizontal dipole requires a 0.5λ element whereas an equivalent vertical dipole only needs a 0.25λ element because the ground plane effectively "images" the other half. Furthermore, vertically polarised antennas generally produce the lower take-off angles required for long range communication systems (>2000 km). However, this take-off angle is more difficult to control as, unlike the horizontal dipole, the height at which a vertical monopole is mounted cannot be varied to achieve the range of elevation angles desired.

The conductive properties of the ground beneath an antenna also play an important role in the resulting radiation pattern. In the South Pole, the ground is predominantly ice, which is a lossy medium. The use of a vertical dipole would thus likely result in poor gain characteristics and a fair amount of power loss. Ground radials of about 0.25λ can be placed directly underneath the vertical monopole to improve efficiency. However, snow accumulation would make them difficult to remove after the project terminates. An advantage of using a vertically polarised transmitter antenna is that its direct radiation would not be detected by the nearby horizontally polarised SuperDARN antenna. Minimal interference with the existing South Pole antenna was a primary design goal, so this especially motivates the use of vertical polarisation over horizontal polarisation.

Horizontal polarisation is preferred for short to medium ranges (<2000 km) because the ground below the antenna reflects the signals upwards and contributes to a radiation pattern with a higher elevation angle and gain. One of the main criteria that the chosen antenna design needs to meet is that it should be able to hold up in the Antarctic environment. This imposes a potential problem with horizontal polarisation because the long antenna elements generally have to be suspended at least 0.25λ above the ground to perform desirably. The radiation pattern of a horizontally polarised antenna in the elevation plane is strongly affected by the presence of the ground underneath. This is because the total radiated signal is the sum of the directly radiated signal and the signal reflected from the ground. The relative phase of these components changes according to height above the ground, resulting in either constructive or destructive interference with the field of the directly radiated rays. Consequently, the elevation radiation angle of horizontally polarised antennas can usually be varied by changing the mounting height.

All of the above being said, the HF transmitter antenna has already been specified by the user to be of horizontal polarisation. Therefore only horizontally polarised antennas were taken into consideration.

3.3 The Impact of the Surrounding Environment

The environment in which an antenna is positioned plays an important role in its performance and design. Especially at large wavelengths, the nearby environment should be as unobstructed as possible. Antennas operating in Antarctica may have their radiation characteristics degraded due to factors such as snow falling on the antenna or building up on the ground. This can particularly affect resonant antennas that have been carefully designed for a single frequency. For a horizontally polarised skywave antenna mounted above the ground, only about half of the signal power travels directly into the sky, while the other half first travels downwards before being reflected back up by the ground to form the final radiation pattern. It is thus vital to have an accurate representation of the dielectric properties of the surrounding ground plane when modelling the antenna.

3.3.1 Investigation into the South Pole Ground

The ground at the South Pole can roughly be broken down into several characteristic layers. From the earth's surface to a depth of approximately 130 m, the ground consists of firn. Firn is a granular type of frozen snow that has not quite been compacted into solid ice yet and has a density in-between that of glacial ice and fresh snow. Below the layer of firn, solid glacial ice then extends to around 2800 m. Beyond that, the ground becomes solid bedrock. A simple breakdown of this is shown in Appendix Figure B.1.

The complex dielectric constant (relative permittivity) of a medium can be described by [41],

$$\epsilon^* = \epsilon_r' - j\epsilon_r''$$ (3.5)

The real part of this constant is known as the ordinary relative permittivity (ϵ_r') and can usually be described by an empirical function of density (ρ) dependent on the medium. According to [42], this function is given for polar ice relative to depth (z) by,

$$\epsilon_r' = (1 + 0.86 \times 10^{-3}\rho(z))^2$$ (3.6)

where $\rho(z)$ is in kg/m^{-3}. This does not strictly apply to all parts of Antarctica as there are many variables involved e.g. temperature profiles, impurity content, age and accumulation [42].

The South Pole, in particular, has a cold and dry ice sheet with low near-surface wind speeds which result in slow ice densification [43]. A depth-density relation has been determined by [44] to be,

$$\rho(z) = \rho_{ice} - (\rho_{ice} - \rho_{surface})e^{\frac{-1.9|z|}{t_f}} \tag{3.7}$$

where $\rho_{ice} = 917\ [kg/m^3]$ is the asymptotic density of polar ice and $\rho_{surface} = 359\ [kg/m^3]$ is the surface density of snow at the South Pole [45]. The firn thickness (t_f) slightly exceeds $100\ m$ at the South Pole [43] but will be set at $100\ m$ for use in the density calculations.

The imaginary term ϵ_r'' is the dielectric loss factor corresponding to the angular frequency (ω). The high frequency conductivity can be related to it by [46],

$$\sigma = \epsilon_r''\epsilon_0\omega \tag{3.8}$$

which allows for the scaling of low frequency kilohertz measurements to megahertz, where dielectric properties become frequency-independent. Figure 3.3 was extracted from [47] and shows the dielectric properties (ϵ_r', ϵ_r'') of an ice core as a function of depth, sampled from Dronning Maud Land on the South Pole Plateau. These measurements were originally taken at 200 kHz but were scaled to a high frequency equivalent of 200 MHz.

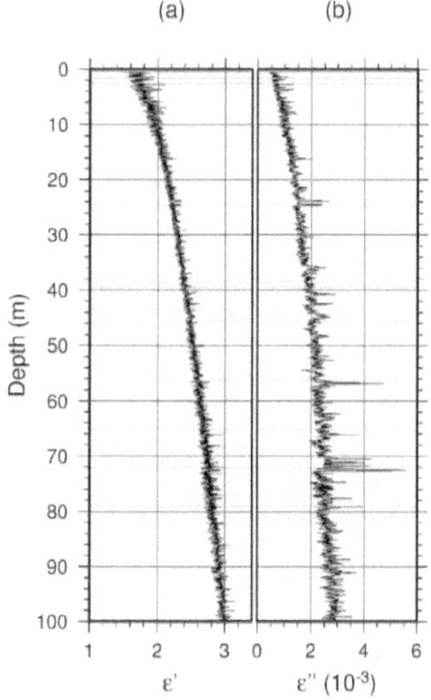

Figure 3.3: Dielectric properties of a South Pole ice core sample extracted from [47] for 200 MHz: (a) Ordinary relative permittivity ϵ_r' and (b) Dielectric loss factor ϵ_r''.

3.4 Suitable Antenna Conductor Materials

A summary of common antenna element materials and some of their electrical and mechanical properties is shown in Table 3.1. The tensile strength (TS) refers to the maximum load-bearing capacity of the material while the modulus of elasticity (E) refers to its resistance to elastic deformation [48]. The relative permeability (μ_r) of all the listed materials is equal to 1. The cost of the materials refers to global prices sourced online as of September 2019 and fluctuates according to the South African Rand exchange rate relative to the US Dollar (1 $USD \approx 14.5\ ZAR$ was used here).

Table 3.1: Typical properties of antenna conductor materials with costs quoted for September 2019.

Material	Conductivity σ (S/m)	Density ρ (kg/m^3)	Modulus of Elasticity E (GPa)	Tensile Strength TS (MPa)	Cost (R/kg)
Copper	6×10^7	8940	110	200	86.77
Aluminium	3.8×10^7	2710	69	90	26.51
Carbon steel	0.6×10^7	7850	207	380	10.50

Copper is the material with the highest conductivity and performance as a radiating element. It is the most resistant to rust and corrosion and is fairly strong. The main shortcoming of copper is its high cost. For antennas that require long solid rods or tubing for the radiating elements, copper is the best option if it can be afforded.

Aluminium is a good conductor but it is also the most malleable. It is therefore less suitable as a long wire in a harsh environment where it will be prone to breakage. Carbon steel is much stronger and cheaper although its conductive properties are poor in comparison to copper. Furthermore Carbon steel contains a high percentage of iron, making it very prone to corrosion in an exposed environment. The best material to select all round would likely be copper-cladded steel. This wire consists of a high grade carbon steel core which has been coated in copper. Copper-cladded steel wire has a tensile strength close to that of plain steel wire but with a conductivity capability between 30% to 40% of that of copper [49]. Therefore, a combination of high conductivity, resistance to corrosion, strength, and low cost would be provided by this type of wire.

The International Annealed Copper Standard (IACS) is a percentage that is often used to quantify the specific conductivity of metals relative to the electrical conductivity of commercially available copper. To get the conductivity (σ) in S/m from the IACS conductivity the following equation can be used:

$$\sigma = \frac{\sigma_{\%IACS} \times 5.8 \times 10^7}{100} \qquad \text{[S/m]} \quad (3.9)$$

3.5 Brief Overview of Directional HF Antennas

For efficient radiation at a given frequency (f), the dimensions of an antenna will conventionally need to be in multiples of half the operating wavelength (λ) determined by,

$$\lambda = \frac{c}{f} \qquad (3.10)$$

A frequency of 12.57 MHz corresponds to 23.87 m, which means that the size of the designed antenna is likely to be substantial.

A horizontally polarised, directional HF transmitter antenna for a medium- to long-range single-hop skywave link is required for the system. A few of the most common antennas that meet these conditions make use of horizontal half-wave dipoles in a log-periodic, curtain array or Yagi-Uda configuration [50]. Other suitable choices are long wire antennas such as the sloping V and horizontal Rhombic [35] [51].

A curtain antenna consists of a matrix of horizontal half-wave dipoles with a reflector screen placed behind the array to make it directional. An example of the smallest array of 2 x 2 dipoles taken from [51] for frequencies above 11 MHz gave total dimensions of over 30 m in both height and width. Similarly, for a 4 x 2 dipole array, dimensions close to 50 m high and 40 m wide are required. Based on the exceptionally large dimensions and the numerous structural components required, the design of a curtain array was discarded from consideration.

Log-periodic dipole arrays (LPDA) and dipole Yagi-Udas are often mistaken for one another because they both consist of slightly different length half-wave dipole elements mounted along a central structural beam. However, there are several important differences between the two. For instance, the LPDA is a broadband antenna whereas the Yagi-Uda is only intended for narrowband applications. In an LPDA, each element is driven 180° out of phase from the adjacent element and the more elements there are, the larger the bandwidth. On the other hand, a Yagi-Uda typically only has one driven element while the rest are parasitic. In the case of the Yagi-Uda, increasing the number of elements increases directionality. Additionally, a Yagi-Uda operating at its resonant frequency will achieve more gain than an LPDA with the same number of elements [52]. The design of a broadband antenna is not regarded as mandatory and so obtaining a higher performance at the single 12.57 MHz frequency takes preference. Consequently, designing a Yagi-Uda antenna over an LPDA was determined to be more appropriate.

The rest of this chapter discusses the operation of the other mentioned applicable antennas in detail. That is to say, the designs of terminated V and Rhombic long wire antennas, as well as various Yagi-Uda configurations, were taken into account. The terminated V and Rhombic can typically be used in more broadband applications, whilst Yagi-Uda antennas are all highly resonant. Even so, a good front-to-back ratio will generally only be achieved at the design frequency.

3.6 Long Wire Travelling Wave Antennas

Travelling wave antennas can be considered transmission-line structures designed to leak radiation. These end-fired antennas are usually associated with being uni-directional and are frequently used in HF applications where the wavelengths are long. Length of radiating elements usually establishes gain and bandwidth, while structure determines the beam shape. In an end-driven long straight wire antenna, the amplitude of the resulting standing wave (symmetric about the wire) will increase with distance along the wire. A main lobe will typically only appreciably develop over several wavelengths long. Furthermore, as the wire's length increases, the resulting beam peaks will approach the axis of the wire. These characteristic beam properties (in relation to the wavelength) are given in Table 3.2, with data extracted from [53].

Table 3.2: Characteristic radiation take-off angles and directivity of a long straight wire antenna.

Length (λ)	Directivity (dB)	Beam peak (°)	Length (λ)	Directivity (dB)	Beam peak (°)
0.5	3.55	64.3	2.5	8.71	30.1
1	5.77	47.2	3	9.30	27.5
1.5	7.06	38.9	3.5	9.81	25.2
2	8.00	33.7	4	10.25	23.8

A standing wave produced by an antenna effectively radiates two waves in opposite directions, one of which is a reflected wave. One way of generating a travelling wave is by terminating a long radiating wire in its characteristic impedance, which effectively reduces the reflected wave. In order to demonstrate this effect, a 4λ long wire driven at a frequency of 12.57 MHz, was simulated in FEKO at a height of 1λ above a perfect electric conductor (PEC) ground plane and is shown in Figure 3.4. These types of antennas are inefficient because half of the transmission power is dissipated into a load. Nonetheless, their simplicity and directive properties make them appealing for HF skywave applications. The two most distinguished terminated long wire antennas for horizontally polarised medium range skywave applications are the sloping V and Rhombic antenna. These will be considered for use in this project and are discussed next.

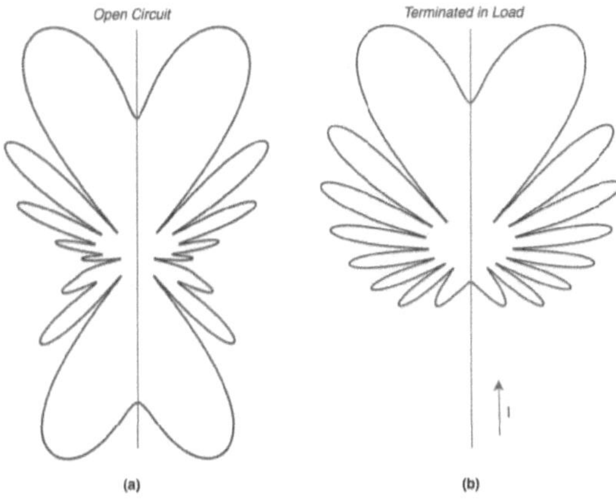

Figure 3.4: Azimuth radiation pattern of a 4λ long wire antenna for: (a) An unterminated standing wave; (b) A terminated travelling wave.

3.6.1 Terminated V Antenna

Consider two equally long wires separated by an apex angle twice that of the corresponding beam peak angles given in Table 3.2, each driven with currents 180° out of phase. In this case, along a plane bisecting the resulting V shape, the beam peaks of the individual wires will add to form a main lobe. In the other directions, the beam peaks will mostly cancel each other out, resulting in a bidirectional antenna. If each wire is then terminated in its characteristic impedance, the beam pattern will form a directional main lobe. This is known as a terminated V antenna. In order to lower the elevation angle of the V directional lobe, the antenna feed point can be elevated with the wires sloping down towards the ground. This is known as a terminated "sloping V" antenna and is shown in Figure 3.5 (a). As the tilt angle approaches the beam peak angle given by Table 3.2, the produced beam will become horizontal. Horizontal polarisation is achieved by feeding the V antenna with a balanced line input.

Problems arise with this type of antenna because the terminating loads need to be connected to a good ground plane. In Antarctica, installing ground radials on the frozen surface may prove problematic, not to mention having to remove them after the experiment is complete. One way of alleviating this is by connecting 0.25λ rods at the ends of the wires. The main appeal of a sloping V is that it is simple to construct, with just two long wires mounted on a single high mast. The sloping V antenna has been implemented for oblique ionospheric sounding previously in a similar project [54]. Some typical parameters were compiled using a combination of sources [55] [53] and are shown in Table 3.3.

Table 3.3: Typical parameter value ranges for a terminated sloping V antenna.

Parameter	Value	Parameter	Value
Leg length (L)	$0.5\lambda - 4\lambda$	Total height of mast (H)	$H_m + H_t$
Feed height (H_m)	$0.5\lambda - 0.75\lambda$	Terminating resistors (R)	$300\Omega - 500\Omega$
Termination height (H_t)	$unknown$	Wire diameter (w_D)	$\approx 0.0001\lambda$

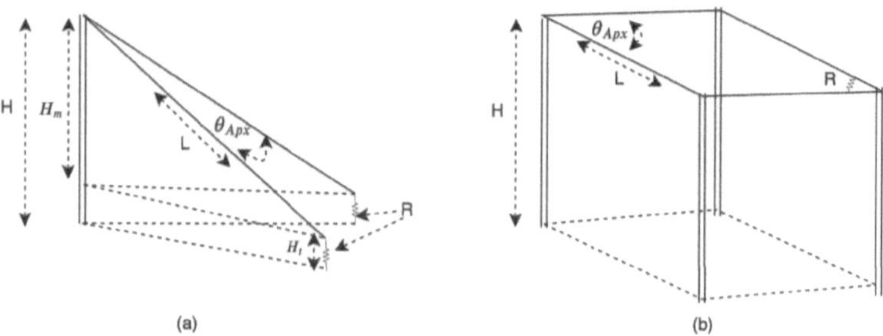

Figure 3.5: Terminated long wire antenna configurations of: (a) A sloping V (b) A horizontal Rhombic.

3.6.2 Terminated Rhombic Antenna

A Rhombic antenna is essentially a combination of two V shaped antennas placed end-to-end and is usually the preferred terminated long wire antenna for HF applications. The structure is shown in Figure 3.5 (b), with apex angles similar to those used by the V antenna for a given element length. The Rhombic configuration has the following advantages over the V configuration according to [55]:

1. For the same total element wire length, slightly more gain can be achieved. For example, a Rhombic antenna with four 2λ elements should have about 1 dBi gain over a V antenna with two 4λ elements.

2. The directional pattern is less frequency sensitive and can be achieved over a slightly wider bandwidth. Accordingly, the apex angle is more variable and can be used to

change the elevation angle of maximum radiation instead of having to slope the antenna (as with the V).

3. The elements are terminated in series by one resistor, which eliminates the need for there to be a good ground plane into which it can terminate.

3.7 Yagi-Uda Antennas

Another method of producing a travelling wave unidirectional pattern is by using the mutual coupling between standing-wave elements in an array. One of the most well known of these is the Yagi-Uda antenna. It makes use of parasitic (non-driven) elements placed in front (directors) and behind (reflectors) a fed element in order to produce an end-fire beam. The parasitic elements are positioned such that they lie in the induction field of the driven element but do not have any electrical connection to the feed. Relative to the driven element, the reflector needs to be slightly longer in order to act like a concave mirror, while the directors need to be slightly shorter in order to focus (converge) the radiation. Although half-wave dipoles and folded dipoles are most commonly used in Yagi-Uda antennas, other resonant elements such as loops can also be used and were considered. Please note that the Yagi-Uda is often simply called a "Yagi" later on in this book for ease of reference.

3.7.1 Horizontal Dipole Yagi-Uda

The conventional geometry of a half-wave dipole Yagi antenna is shown in Figure 3.6. The resonant length (I) of a centre-fed half-wave dipole in the presence of parasitic elements usually needs to be adjusted to around $0.45\lambda - 0.48\lambda$. Although the impedance of a stand-alone dipole is usually around 73Ω in free space, closely spaced parasitic elements typically cause this impedance to fall below 50Ω (standard input feed impedance) resulting in a poor VSWR. One way of increasing this impedance, for matching purposes, is by using a folded dipole as the driver element.

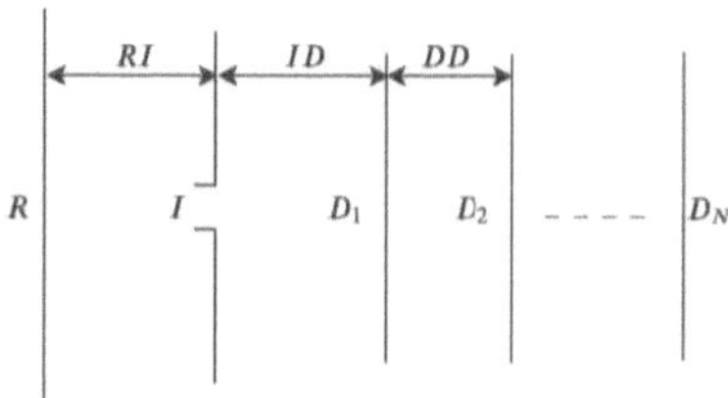

Figure 3.6: Classic half-wave dipole Yagi-Uda antenna configuration.

The reflector length (R) needs to be made approximately 5% longer than the driver element, resulting in it having a slightly lower resonant frequency. Consequently, the reflector element becomes inductive. On the other hand, the director length (D) needs to be made approximately 5% shorter than the driver element, resulting in it having a higher resonant frequency. Accordingly, the director element becomes capacitive. This causes phase distribution across the elements and produces a travelling plane wave.

Generally only one reflector is required and is the most responsible for improving the front-to-back ratio of the antenna. Adding more than one reflector does not really impact the overall antenna performance. On the other hand, there can be numerous directors with their main function being to produce a more focused high gain front lobe. Generally, the length of the driver element and the first director have the most influence on the gain. The diameter of both the wire and mounting beam (if made of conductive material) will have an effect on the optimum element lengths required and therefore should be carefully considered.

Half-wave dipole Yagi antennas have been extensively used and analysed for directional HF applications and the typical design parameters are compiled in Table 3.4. The Yagi's performance can be improved by feeding more than just one element, as demonstrated by the four-element design with a criss-cross "backfire" feed in Figure 3.7 (b). Although there are more elements, the total antenna length required will actually be shorter than that of the standard three-element antenna seen in Figure 3.7 (a). This is because all of the elements will need to be more closely spaced to one another. The main consequence of the dual-fed elements is an improvement in FTB ratio. The simple two-element "backfire" antenna shown in Figure 3.7 (c) is also possible, but some gain and bandwidth will be sacrificed. Table 3.5 shows an example of the comparison in performance between the various setups, which was compiled from findings in [53].

Table 3.4: Typical parameter value ranges for a half-wave dipole Yagi-Uda antenna.

Parameter	Value	Parameter	Value
Driver length (I)	$0.45\lambda - 0.48\lambda$	Reflector-Driver (RI)	$0.1\lambda - 0.25\lambda$
Reflector length (R)	$0.47\lambda - 0.5\lambda$	Director-Director (DD)	$0.1\lambda - 0.5\lambda$
Director length(s) (D)	$0.42\lambda - 0.45\lambda$	Wire diameter (w_D)	$0.001\lambda - 0.0085\lambda$

It is is also possible to stack Yagis in order to achieve additional gain and a more focused beamwidth [56]. However, it was decided that this would unnecessarily complicate construction and was therefore not entertained as a possibility.

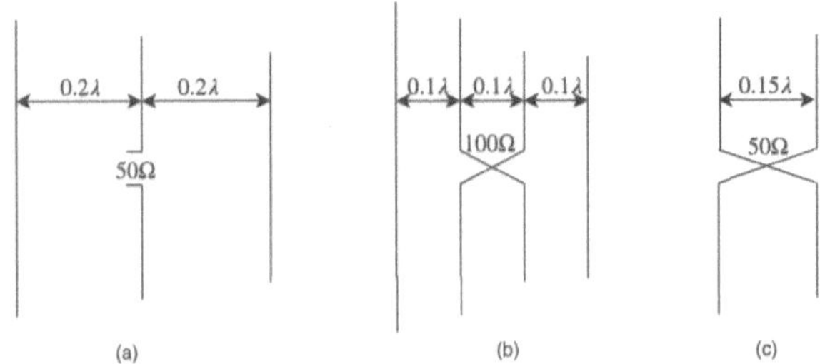

Figure 3.7: Alternate half-wave dipole Yagi-Uda configurations for a wire diameter of 0.006λ.

3.7.2 Resonant Loop Yagi-Uda

As already mentioned, the elements in a Yagi antenna do not have to be half-wave dipoles. Instead, resonant loop elements can be utilised. Resonance in a single loop element characteristically occurs at a perimeter of 1λ and the maximum signal radiated will be normal to the plane in which the loop lies. Horizontal polarisation is produced when the loop is

Table 3.5: Performance comparison of a few optimised half-wave dipole Yagi-Uda configurations compiled from findings in [53].

Configuration	Gain (dB)	F/B (dB)	VSWR
2 Element dual-feed (25Ω source)	5.9	34.1	1.61
3 Element single-feed (50Ω source)	7.6	18.6	1.57
4 Element dual-feed (50Ω source)	7	53	1.25
6 Element single-feed (50Ω source)	9.7	22.7	1.61

fed across the bottom of the element. Similarly, vertical polarisation is produced when it is fed on the length of element perpendicular to the ground. The two polarisation layouts are shown in Figure 3.8.

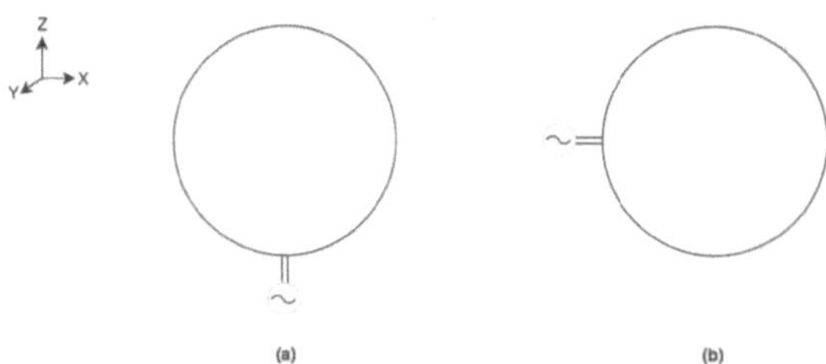

Figure 3.8: Orientation of a loop element feed point to produce: (a) Horizontal polarisation; (b) Vertical polarisation.

A single loop above a ground plane has a bidirectional radiation pattern similar to that of a dipole. This was demonstrated in FEKO by simulating a full-wave loop and half-wave dipole at 0.5λ above a PEC ground plane. The resulting traces in the elevation plane (side view) are shown in Figure 3.9. It can be observed that the loop element radiates a bit less towards the sky than the dipole, making it marginally better suited to applications requiring low take-off angles. Understandably, the loop achieves slightly more maximum gain owing to the fact that it is 1λ long as opposed to 0.5λ. A Yagi can, in fact, be designed with a combination of both dipole and loop elements. However, this once again imposes additional construction complexity for seemingly little benefit and was thus disregarded.

The loop can be any closed geometric shape, although the choice of shape will have a slight effect on input impedance and the resonant perimeter. Owing to sinusoidal current distribution, the current on the loop will reach a maximum (virtual short-circuit point) at about 0.5λ from the feed point [53]. Correspondingly, the maximum standing-wave voltage will occur 90$°$ out of phase at points 0.25λ from where the loop is fed. The loop shape should be designed such that this point lies in the middle of a segment and does not fall on a discontinuity (such as an edge) where air breakdown can occur. Standard loop element shapes include quadrilaterals, hexagons, triangles and circles. Quadrilateral shapes are mostly preferred at large wavelengths as they can easily be constructed by stretching wires over a supporting frame (such as an X frame), or run inside a protective

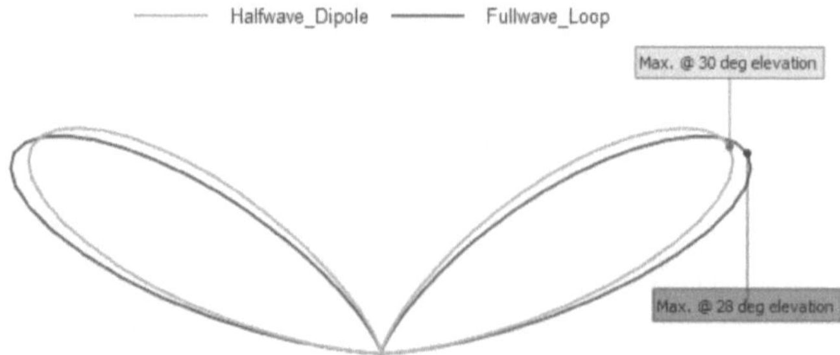

Figure 3.9: Simulated radiation elevation traces of a half-wave dipole and full-wave loop positioned 0.5λ above a PEC ground for horizontal polarisation.

shell made of non-conducting material.

Recall that a Yagi consisting of half-wave dipoles will normally require a driver element shorter than the resonant length (0.5λ) of a single half-wave dipole. On the contrary, a Yagi with loop elements will most often necessitate a driver that is slightly longer than the resonant length (1λ) of a single loop. Once again, the reflector element is generally 5% longer than the driver element while any subsequent directors are at least 5% smaller. Element spacing lies between $0.15\lambda - 0.2\lambda$. If preferred, it is possible to keep all elements equal in length. This can be done by placing small electrical stubs on the bottom of the elements to effectively increase or decrease their electrical size. The reflector element would require a short-circuit stub in order to make it inductive, while directors would require open-circuit stubs to make them capacitive. The geometry of a quadrilateral loop Yagi is shown in Figure 3.10.

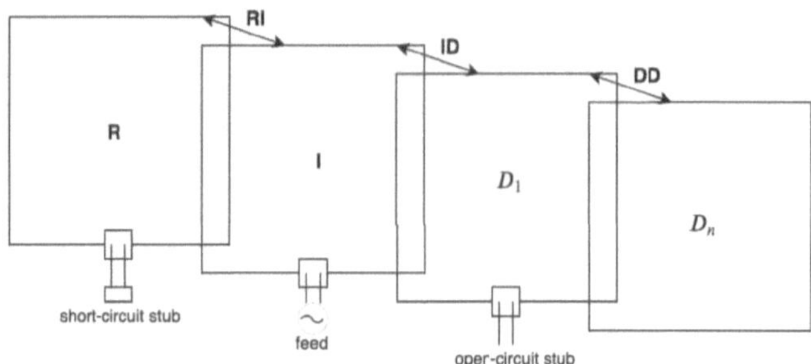

Figure 3.10: Classic loop Yagi-Uda antenna configuration for horizontal polarisation.

It has been reported that a rectangular loop twice as long in length than width will provide slightly more gain than its square counterpart. A rectangular shape is also appealing because, if it is mounted with the longest sides parallel to the ground, the total height to which the antenna extends will be reduced. Unlike dipole elements, the symmetry of a loop about an axis reduces the effect of the mounting beam structure on the antenna performance. Hence, a central conductive metal rod can be used with little resulting interference.

3.8   Chapter Summary

The main intention of this chapter was to outline significant basic antenna design aspects in order to enable the selection of appropriate HF antennas for review. The importance of antenna polarisation and the ground plane were discussed in depth. It was found that vertically polarised antennas could actually be viable as, in skywave applications, the polarisation of the transmitter and receiver does not actually need to match. Transmitting vertically polarised waves would contribute minimal interference to the South Pole SuperDARN array. Nonetheless, there are other beneficial reasons for using horizontal polarisation. Thus, as it was already specified by the user, only horizontally polarised antennas were considered.

The important dielectric properties of the ground medium at the South Pole were examined. Specific data was extracted from the published findings of an experiment using South Pole ice core samples. This data was used later on to model an antenna ground plane for simulations. Additionally, suitable conductor materials were assessed and it was decided that either copper or copper-clad steel antenna elements would be the most practical.

Based on the research done, a few suitable antennas were chosen for further investigation. The terminated sloping V and Rhombic long wire antennas were selected because of their apparently simple and robust design. Common Yagi antenna configurations utilising single- or dual-fed half-wave horizontal dipoles were also considered, as well as those utilising full-wave loop elements. Further deliberation on these antennas is given in the design and simulation chapters.

4 Ionospheric Path Analysis and Link Budget

The main purpose of this book was to design an antenna with a radiation pattern and gain such that a point-to-point skywave link can e ectively be established, while transmitting as little power as possible. Since a transmitter has already been designed for a set operating frequency of 12.57 MHz and maximum power output of 1W, the next most important criterion are the elevation angle of radiation and the antenna gain required. Thus an in depth link analysis needed to be done in order to quantify the losses expected and to better understand the calibre of the system that can be achieved.

This chapter starts by detailing the method and results of the analysis done using propagation models to determine the path availability and range of elevation angles needed for transmitted signals to reach the receiver. The chapter goes on to define some of the transmission link losses a skywave radio wave will typically encounter and outlines the properties of the SuperDARN receiver. The total ionospheric absorption loss predicted by ICEPAC was considered, following upon which a manual calculation of the specific non-deviative absorption that may be seen was done using the magnetoionic theory discussed in the earlier sections. This was done in order to supplement ICEPAC estimates and ensure that enough power is delivered to both the ordinary and commonly neglected extraordinary wave modes. The further investigation into the D/E region absorption also allowed for a brief examination of the validity of the utilised models in the Polar region. Other prevalent link losses were considered and, combining all of the information obtained, a complete link budget was compiled.

4.1 Path Availability Analysis

As previously discussed, solar activity is the biggest contributor to variations in the ionosphere. The ray path between the South Pole and SANAE at di erent stages of an 11 year solar cycle therefore needed to be examined. We are currently at the end of solar cycle 24 with a solar minimum expected to occur in 2020. A decision was taken to fo-cus all analyses on solar cycle 23 [1997-2008] as it is the most recent fully completed cycle.

A brief investigation into the most probable conditions for open communication was done using the online VT ray tracing tool. This interface requires one to select the SuperDARN radar under consideration, beam number, frequency, as well as a time and date. The ray tracer outputs are relative to transmission from the SuperDARN, but the computed rays can be considered to be reciprocal due to the simplified ray theory used. The maximum ground range computed by the tracer is 2000 km, which is the approximate distance between SANAE and the South Pole. Beam number 5 was selected as it aligns the most closely with the South Pole (Figure 4.1). The exact operational frequency of 12.57 MHz cannot be selected and a choice between 12 MHz or 13 MHz had to be made. Using 12 MHz will output a higher refractive index, thus increasing the chances of the communication link being declared open. Using 13 MHz is more conservative, but a closed path could be anticipated when it is in fact open. Appendix Figure A.1 shows an example of the di erence in the paths calculated for a 12 MHz and 13 MHz ray. In the analysis done, both frequencies were taken into account and a median result taken.

The ray tracer clearly indicated that one-hop propagation was the most probable route through the ionosphere for all seasons, times of day and points in the solar cycle. Al-

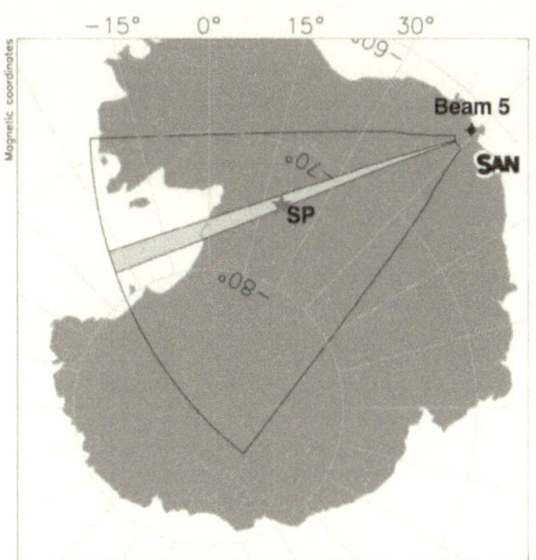

Figure 4.1: Beam 5 of SANAE SuperDARN in alignment with the geographic South Pole

though at times of elevated ionospheric electron density, such as summer during solar maximum, 2-hop propagation is also possible. However, the amount of scatter received via this mode is minimal in any case and was therefore not considered.

4.1.1 Results

Using the VT ray tracing tool, the relative power densities of signal ground scatter arriving within 100 km of the 2000 km ground range mark were evaluated for both midday (12:00 UT) and midnight (00:00 UT). Declaring whether the path was open or not was simplified into three categories, namely,

o : *Path open, strong relative signal strength.*
∼ : *Path intermittent, weak relative signal strength.*
x : *Path closed, within 100 km of receiver.*

Utilising the key described above, Table 4.1 was compiled. An example of the power density profile generated by the VT program can be seen in Appendix Figure A.2. In these power density profiles, a relative power of less than -15 dB was considered weak.

4.1.2 Discussion

It can quickly be observed that during the peak winter months of June and July, the path was almost never open. This is because in the polar cap region, at the March equinox, the sun sets at approximately 12:00 UT and does not rise again until the September equinox. The levels of solar-driven ionisation therefore steadily decrease to a minimum around mid winter, which results in the rays not being sufficiently refracted back down to the receiver. An example of this is depicted in Figure 4.2. However, in June 2001, because of the increased solar activity approaching the solar maximum of November 2001, it was found that intermittent signal was in fact received on a few afternoons.

Table 4.1: Path availability of 12.57 MHz ray at various points of solar cycle 23

Cycle 23	Start: 1997 (post min)		1999		2001 (pre max)		2002 (post max)		2005		End: 2008 (pre min)	
UT/ Month	12:00	00:00	12:00	00:00	12:00	00:00	12:00	00:00	12:00	00:00	12:00	00:00
Jan	~	o	o	o	o	o	o	o	o	o	x	~
Feb	x	x	o	o	o	o	o	o	~	x	x	x
Mar	x	x	o	x	o	x	o	x	x	x	x	x
Apr	x	x	~	x	~	x	~	x	x	x	x	x
May	x	x	~	x	~	x	~	x	x	x	x	x
Jun	x	x	x	x	~	x	x	x	x	x	x	x
Jul	x	x	x	x	x	x	x	x	x	x	x	x
Aug	x	x	~	x	~	x	~	x	x	x	x	x
Sep	x	x	~	x	~	x	~	x	x	x	x	x
Oct	~	x	o	o	o	o	o	o	~	x	x	x
Nov	o	o	o	o	o	o	o	o	o	o	x	x
Dec	o	o	o	o	o	o	o	o	o	o	x	~

At any stage of the solar cycle consistent communication is most probable during the months of November, December and January. Approaching the solar minimum that occurred in December 2008, even in peak summer, the chances of communication were very low. It appears that any possible open communication close to the solar minimum was more likely to happen at midnight in summer than at midday. This agrees with the higher ionisation levels observed at 00:00 UT rather than 12:00 UT in the electron density profiles previously generated from the IRI and shown in Figure 2.1.

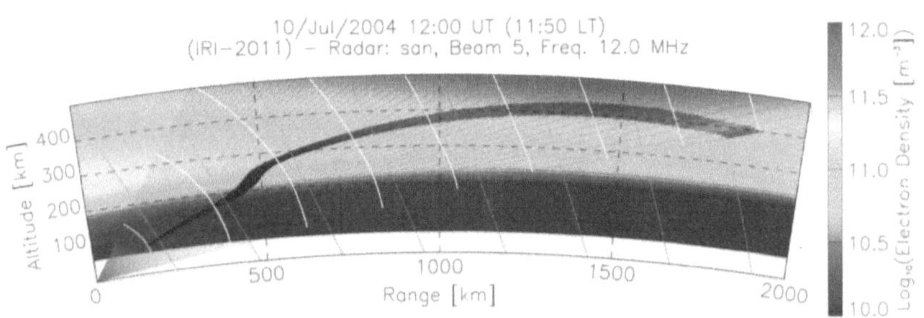

Figure 4.2: Example of the insufficient refraction that occurs during winter in Antarctica generated using the VT ray tracer

Following on from the year halfway between solar minimum and solar maximum (1999), an open path could be achieved almost 24 hours a day for 5 months annually. Intermittent signal could also be received in the autumn and spring months near to the equinoxes, especially during the afternoons. On the other hand, following on from the year halfway between solar maximum and solar minimum (2005), path availability had already deteriorated significantly and only January, November and December were found to have reliable communication.

This leads to the conclusion that the system should be optimised for operation in years nearest to solar maximum, particularly during the spring and summer months. Unfortunately, because of the higher levels of ionisation and electron density in the ionosphere during these times, this also means that the wave will typically see the most loss and will thus require higher power supplied to the transmitted wave. January, November and December halfway to solar minimum (2005) were taken as the last points at which the designed system should still theoretically be able to attain communication.

4.2   Elevation Angle Analysis

Considering that the ICEPAC propagation prediction program has some high latitude capabilities, it was used to approximate the elevation angles of radiation required for the signal to reach the receiver under various conditions. It is recommended by the ICEPAC developers that the Smoothed Sunspot Number (SSN) listed by the National Geophysical Data Centre be utilised as these are the values to which ICEPAC has been calibrated. A copy of this data can be seen in Figure A.3 in the Appendix. The default CCIR (Oslo) ionospheric coefficients were selected. The CCIR (Oslo) coefficients use monthly median values, so a particular day of the month cannot be specified. The analysis was done for a year in which solar maximum occurred (2001), and for another year half way to solar minimum (2005). Recall that the path will not necessarily be open at all approaching a solar minimum. Therefore a halfway point was considered the limiting factor on the bracket of elevation angles necessary for a functioning system.

4.2.1   Results

A graph showing the elevation angles predicted every 2 hours for each month in 2001 is shown in Figure 4.3. Polynomial lines of the 6th order were fitted to the data to observe the general trend. Predictions for 2005 were done for six months of the year, although only three of the months being January, November and December, actually had an open path . The data retrieved from ICEPAC for 2001 and 2005 can be seen in Table A.1 and Table A.2 in the Appendix respectively. The average elevation angles taken from the data for select months are shown in Table 4.2.

Table 4.2: Average monthly elevation angles predicted by ICEPAC

Year	January	February	March	October	November	December
2001	18.58	14.55	13.44	14.09	17.69	18.64
2005	16.04	14.39	12.99	12.57	15.47	15.35

4.2.2   Discussion

Table 4.2 shows that, on average, higher elevation angles are required in summer and years tending to solar maximum than in winter and years tending to solar minimum. This is because lower elevation angles will be refracted very steeply in regions of high electron density which could result in the ray falling short of the receiver. Figure 2.1 showed that the peak F2 electron density was smaller at 12:00 UT than at 00:00 UT and occurred at a lower altitude. As confirmed in Figure 4.3, transmitting a slightly lower angle at 00:00 UT than at 12:00 UT is then necessary because the ray will have to travel slightly further before reaching the higher reflection layer where it will rapidly be refracted back down.

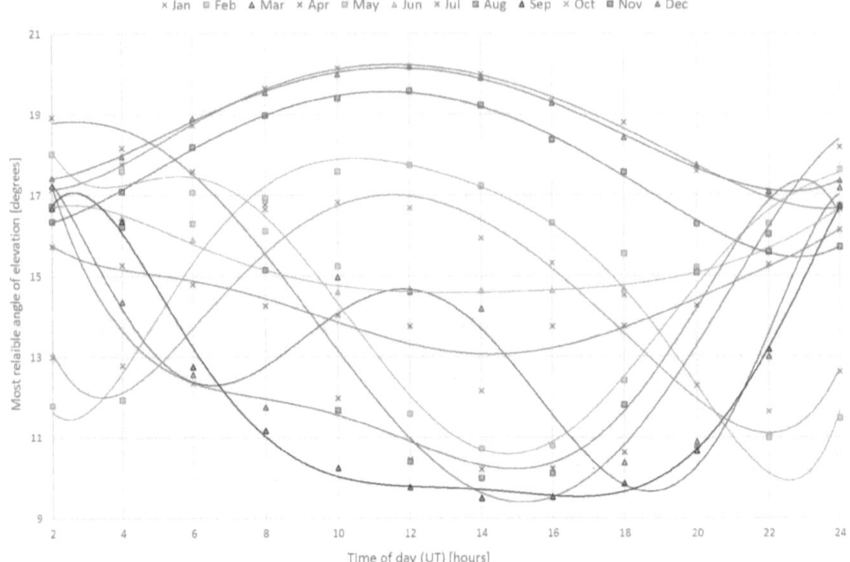

Figure 4.3: Hourly elevation angles predicted by ICEPAC for the months in 2001.

Once again, referring back to the electron density of Figure 2.1, during winter near the solar maximum of 2001 the peak F2 electron density was higher at 12:00 UT than at 00:00 UT and occurred at a lower altitude. A lower elevation angle at 12:00 UT would then be needed to reach the receiver than at 00:00 UT because it will take slightly longer to reach the lower layer but be refracted more strongly due to the higher oblique incidence and electron density. Considering that the system design is only for operation within the months that the path will most reliably be open (January, February, October, November and December), winter month elevation angles are of little interest.

Within the operational months of both 2001 and 2005, the lowest angle of elevation predicted by ICEPAC was in February 2001 and was 10.78° at 20:00 UT. The average angle of that same month was 14.55°. The lowest average angle was found in March of 2001 at 13.44°, but considering that the path at this time will only actually be open for a few hours of the afternoon, a slightly higher value can be taken.

For the most part, the most valuable months of system operations will be November through to January. Thus an antenna with enough gain at a minimum elevation angle of 14° should suffice, although it would be beneficial to get this angle even lower.

4.3 Transmission Loss

The system transmission power loss (L_{tr}) that a radio wave will experience from transmitter to receiver (excluding hardware losses) can be summarised by,

$$L_{tr} = L_{fs} + L_i + L_{gr} + L_{pol} - (G_t + G_r) \qquad \text{[dB]} \qquad (4.1)$$

with G_t and G_r being the gains (relative to an isotropic antenna in free space) of the transmitter and receiver antennas respectively. The term L_{gr} represents the ground reflection losses stemming from multiple ray hops. However, only a single ray hop was considered here, so ground reflection losses were neglected.

The largest loss factor is the basic free space loss (L_{fs}) caused by geometrical spreading of energy as the radio wave gets further away from the transmitter [29]. This loss is estimated using,

$$L_{fs} = 32.45 + 20\log f + 20\log D \qquad (4.2)$$

with the frequency of the wave (f) in MHz and the ground range distance (D) in kilometres. This equation was derived based on a simple model in which the earth and ionosphere are considered flat. This gave a free space loss of approximately 121 dB.

As previously discussed, a signal travelling through the ionosphere will become elliptically polarised. Polarisation mismatch loss (L_{pol}) is caused by the received signal polarisation not being perfectly matched to the antenna polarisation. This depolarisation can result in a half power loss of 3 dB in a worst case scenario for skywave propagation. Polarisation loss of such is implicitly included in ICEPAC predictions because the ionospheric constants used in computations were obtained through real measurements inclusive of such losses.

The total ionospheric absorption loss (L_i) is a combination of the non-deviative absorption (L_a) in the D and E layers, as well as deviative losses, over-the-MUF losses and sporadic E losses. Estimation of the total ionospheric absorption loss is discussed in detail in the next sections.

4.4 Ionospheric Absorption Predicted by ICEPAC

ICEPAC estimates the non-deviative absorption of the oblique ordinary wave by a lumped empirical equation related to an absorption index (I), which includes the effects of solar activity variation and zenith angle. The absorption losses above the non-deviative absorption are represented as an excess system loss (with auroral losses included). This excess system loss is based on estimates of uncertainties which were obtained by comparing actual received signal powers with the predictions of received signal powers [29]. ICEPAC outputs a total ionospheric absorption (L_i) which includes all of these losses in one lump sum. However, the majority of this quoted loss will be non-deviative absorption.

4.4.1 Results

The absorption under normal conditions will peak during the summer of solar maximum as this is when the greatest levels of ionisation occur in the various ionospheric layers. Therefore, the total ionospheric absorption losses seen on the predicted one-hop propagation path were generated for November and December 2001, and for January 2002. December of 2005 was also generated in order to see the effect of a lower sunspot number. There was no focus on the absorption at solar minimum, as it has already been determined that the path will hardly be open. The results are shown in Table 4.3.

Table 4.3: Hourly ionospheric absorption loss estimated by ICEPAC for key months

| Time (UT) | Ionospheric Absorption L_i (dB) | | | | Time (UT) | Ionospheric Absorption L_i (dB) | | | |
	Nov 2001	Dec 2001	Jan 2002	Dec 2005		Nov 2001	Dec 2001	Jan 2002	Dec 2005
2:00	2.80	4.16	2.96	3.28	14:00	5.20	7.22	5.40	5.92
4:00	3.28	4.93	3.42	3.88	16:00	4.71	6.66	4.82	5.42
6:00	4.01	5.96	4.26	4.67	18:00	3.93	5.79	4.04	4.63
8:00	4.83	6.98	5.09	5.48	20:00	3.25	4.70	3.35	3.79
10:00	5.46	7.74	5.58	6.03	22:00	2.85	4.10	2.90	3.25
12:00	5.53	7.76	5.74	6.16	24:00	2.69	3.89	2.84	3.05

4.4.2 Discussion

A peak ionospheric absorption loss of 7.76 dB was predicted for 12:00 UT in December 2001. This was expected as it was midsummer near the solar maximum which occurred in November 2001. Around 1.6 dB less absorption was predicted for the same month and time in 2005. The absorption for the equinox month, March of 2001, was found to only be a maximum of 3.86 dB at 12:00 UT. Owing to the fact that the worst case scenario absorption under normal conditions was being investigated for use in the link budget, December 2001 was chiefly scrutinised in further investigations done.

4.5 Manual Non-Deviative Absorption Calculations

The non-deviative absorption (L_a) in the D and lower E layers alone is the second largest loss factor after free space loss and required an in depth analysis in order to account for the effects of the geomagnetic field. As previously mentioned, ICEPAC only considers absorption of the ordinary wave mode. Since the non-deviative absorption of the extraordinary mode will be higher than that of the ordinary mode, an additional loss needs to be quantified so as to ensure that enough power is also delivered to the extraordinary mode. Furthermore, ICEPAC uses a lumped equation primarily based on observations in the Northern Hemisphere and does not output the non-deviative absorption individually. For this reason, manual non-deviative absorption calculations were done so as to include a better informed estimate of the ionospheric loss for use in the link budget.

4.5.1 Method and Assumptions

Ignoring the gyrofrequency and using Equation (2.26), a typical absorption-height profile was generated for a 12.57 MHz wave *vertically* incident at a midpoint between the South Pole and SANAE station. The results of which can be seen in Figure 4.4 and correspond to a geographic location of $80.833°S$ and $2.7802°W$. Electron densities were retrieved from the IRI model, while total electron collision frequencies were calculated from Equations (2.8) to (2.13) where surrounding neutral particle densities were taken from the NRLMSISE-00 model. All of the models used describe the ionosphere vertically in minimum increments of 1 km for specified geographical coordinates and dates.

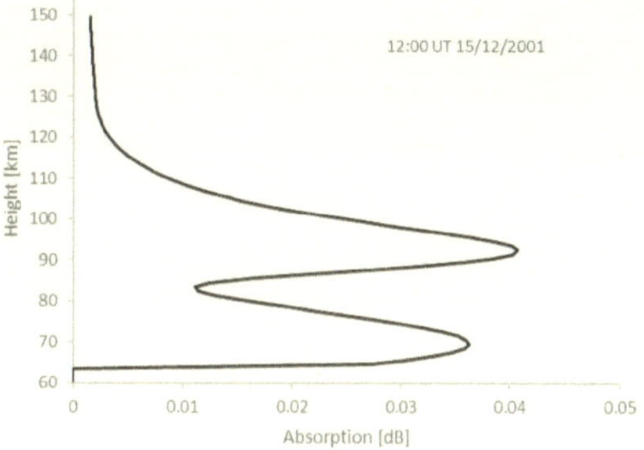

Figure 4.4: Non-deviative absorption height profile generated for a midpoint between the South Pole and SANAE station

It can be seen from Figure 4.4 that, above a height of 110 km, the absorption per kilometre starts to fall below 0.01 dB. Under 65 km, the IRI model returns an electron density of zero, whilst the cumulative total from 110km to 150km only results in an additional 0.1 dB loss (or effectively 0.2 dB for both entry and exit). Accordingly, for all ensuing calculations done, the height limits of the absorption integral were set from 65km to 110km.

A one-hop oblique incidence wave will traverse the absorptive region of the ionosphere twice i.e. once on entry and once on exit. Two separate absorption profiles for different locations thus needed to be generated and added together to find an estimate of the total D/E layer absorption. In reality, the angle of D layer entry (ϕ_i) is not exactly equal to the returning exit angle of the ray, but symmetry was assumed for the purpose of ray tracing and any calculations done.

The geographic coordinates corresponding to the most likely points of ray entry and exit from the ionosphere had to be estimated in order to retrieve relevant data from the models. Using the angle of elevation (θ_{el}) of the ray, both the angle of oblique incidence (ϕ_i) and the ground range (L) from the transmitter can be found for a height of ionospheric entry (h_i). Given the geometry shown in Figure 4.5, the following relations were thus found, relative to the radius of the earth (R),

$$\phi_i = \arcsin\left(\frac{R\cos\theta_{el}}{R+h_i}\right) \tag{4.3}$$

$$L = R \times \left[\frac{\pi}{2} - \theta_{el} - \arcsin\left(\frac{R\cos\theta_{el}}{R+h_i}\right)\right] \tag{4.4}$$

The midpoint height of the defined integral region (65 km to 110 km) was selected for use in the equations i.e. $h_i = 85\ km$. Using the relations determined, the geographic entry/exit co-ordinates were estimated along a straight path between the South Pole and SANAE for different angles of elevation. Table 4.4 shows the resulting exit and entry coordinates for a few key elevation angles. These were estimated by assuming that the ground range from the transmitter to point of ray entry was approximately the same as the ground range from the receiver to point of ray exit.

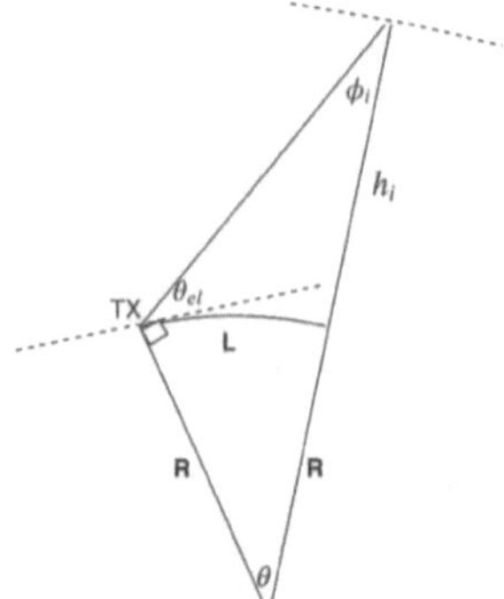

Figure 4.5: Geometry of parameters used in absorption calculations

Table 4.4: Estimated geographic coordinates of an obliquely incident ray quoted in decimal degrees

Elevation Angle θ_{el} (°)	Ground Range L (km)	Entry Coordinates		Exit Coordinates	
		Latitude	Longitude	Latitude	Longitude
13.75	315	-87.1566	357.2358	-74.51	357.2036
15.5	280	-87.4833	357.2674	-74.1957	357.1961
18	245	-87.81	357.299	-73.8813	357.1886
21	210	-88.1448	357.3194	-73.5668	357.19
25	175	-88.4796	357.3398	-73.2522	357.1913

Height-dependent year long magnetic field flux densities (B) for the given coordinates were obtained from the IGRF model and applied to Equation (2.7) to extract respective gyrofrequencies. The gyrofrequencies over the integral range did not vary much. Hence an average gyrofrequency was applied in calculations. The IGRF model was also used to determine the geomagnetic field's angle of inclination, namely the magnetic dip (Φ_{dip}). Calculations required the taking of the vertical absorption of a wave for an equivalent frequency and translating it to a corresponding oblique absorption. Owing to this, the angle θ_H needed in the calculations was simply the angle between the magnetic field and the vertical i.e. $90° - \Phi_{dip}$. Figure 4.6 shows a simple diagram summary (not drawn to scale) of all the relevant parameters mentioned.

ICEPAC predictions confirmed that absorption will be greatest in the summer months, especially at solar maximum. Estimating the absorption under these conditions was thus the focus of most of the calculations done. The best angle of elevation for ray takeoff varies hourly, so for the purpose of straightforward analyses, only the average elevation angles predicted by ICEPAC for the particular month under consideration were applied to each scenario.

All the results reported include the equivalent vertical wave absorption calculated before being translated across to the equivalent oblique wave absorption. Both extraordi-

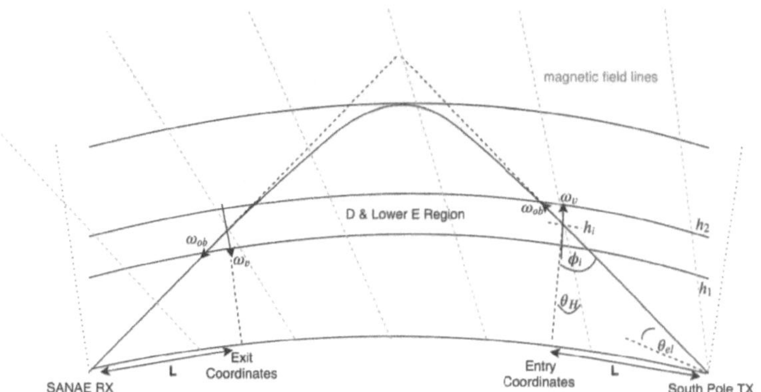

Figure 4.6: Summary of simple ray model parameters used in absorption calculations

nary and ordinary mode absorption were calculated. The results of simply omitting the gyrofrequency were also shown.

4.5.2 Results

Employing all of the assumptions made along with equation (2.29), the following absorption calculations were done and analysed for a 12.57 MHz obliquely incident wave:

1. Every 2 hours of a full day in mid summer following solar maximum. The day chosen was 30/12/2001 with an elevation angle of 18°. This was done to confirm the approximate time of peak absorption during summer and the results can be seen in the graph in Figure 4.7. This was also done for the actual solar maximum during November on 6/11/2001 at 15.5° elevation and is shown in Figure A.5 of the Appendix.

2. For the peak time on different days over the summer months following a solar maximum. This was done to determine the approximate date at which summer absorption peaks. The dates selected were 10/12/2001, 20/12/2001, 30/12/2001, 10/01/2002 and 20/01/2002 at 12:00 UT with elevation angle 18°. The results can be seen in Table 4.5.

3. For different elevation angles (13.75°, 18° and 25°) on the same day 30/12/2001 at the peak time 12:00 UT shown in Figure 4.7. The results are shown in Table 4.6. In this case only, the data from the same geographic location was used for the various elevation angles for the purpose of a fair comparison.

4. For mid solar cycle (leading up to solar minimum) and mid summer at 12:00 UT. The date selected was 30/12/2005 at elevation 15.5°. The results of this were shown as separate entry and exit absorption so as to demonstrate the geographic variance in the calculations. Table 4.7 shows this data.

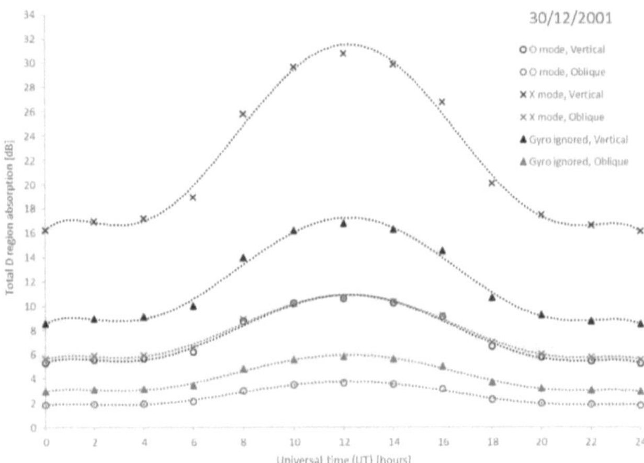

Figure 4.7: Hourly non-deviative absorption profiles computed using defined assumptions for December 2001.

Table 4.5: Total non-deviative absorption calculated at 12:00 UT for an elevation angle of 18°.

Date dd/mm/yyyy 12:00 UT	O mode Vertical (dB)	O mode Oblique (dB)	X mode Vertical (dB)	X mode Oblique (dB)	Gyro ignored Vertical (dB)	Gyro ignored Oblique (dB)
10/12/2001	11.24	3.88	31.43	10.85	17.52	6.05
20/12/2001	11.29	3.90	31.49	10.87	17.58	6.07
30/12/2001	10.61	3.66	30.76	10.62	16.79	5.80
10/01/2002	10.21	3.52	28.23	9.74	15.84	5.47
20/01/2002	9.36	3.23	25.76	8.89	14.50	5.00

Table 4.6: Total non-deviative absorption calculated for different elevation angles corresponding to 12:00 UT on 30/12/2001.

Elevation angle (°)	O mode Vertical (dB)	O mode Oblique (dB)	X mode Vertical (dB)	X mode Oblique (dB)	Gyro ignored Vertical (dB)	Gyro ignored Oblique (dB)
13.75	14.43	4.11	50.62	14.42	24.34	6.94
18	10.61	3.66	30.76	10.62	16.79	5.80
25	7.11	3.18	15.67	7.00	10.13	4.53

4.5.3 Discussion

The absorption profiles of Figure 4.7 confirmed that summertime absorption typically peaks daily at about 12:00 UT. Table 4.5 indicates that the date of peak absorption lies around the 20th of December. Accordingly, at 12:00 UT on the 30/12/2001 for an elevation angle of 18°, it was found that the ordinary oblique absorption was 3.9 dB, while the extraordinary oblique absorption was 10.87 dB. When the effects of the magnetic field were ignored (i.e. $\omega_H = 0$), the absorption seen was 6.07 dB. It will be recalled that the vertical wave absorption depicted was calculated for a lower equivalent frequency than that of the oblique wave and is consequently always larger. This compensates for the fact

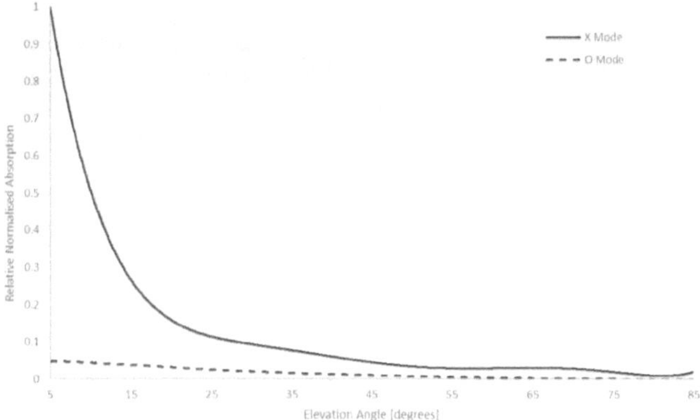

Figure 4.8: Normalised difference in the absorption of extraordinary and ordinary modes relative to the take-off angle of the wave.

Table 4.7: Separate entry and exit non-deviative absorption calculated at 12:00 UT on 30/12/2005 for an elevation angle of 15.5°.

Coordinates	O mode Vertical (dB)	O mode Oblique (dB)	X mode Vertical (dB)	X mode Oblique (dB)	Gyro ignored Vertical (dB)	Gyro ignored Oblique (dB)
Entry						
Lat: -87.4833	3.10	0.96	12.20	3.78	5.47	1.69
Lon: 357.2674						
Exit						
Lat: -74.1957	4.61	1.43	13.44	4.16	7.33	2.27
Lon: 357.196						
Total Absorption	**7.70**	**2.38**	**25.64**	**7.93**	**12.80**	**3.96**

that the equivalent vertical wave would technically travel a smaller distance through the ionosphere.

As elevation angle (θ_{el}) decreases, incident angle (ϕ_i) increases and the wave becomes more oblique. It can be seen from Table 4.6 that the more oblique the wave, the greater the absorption. This is because the more oblique wave will have to travel a greater distance through the absorptive ionospheric region. At an elevation angle of 13.75°, a notable extraordinary mode loss of 14.42 dB was determined. Figure 4.8 shows the normalised absorption of the extraordinary and ordinary mode as the elevation angle varies. As can be seen, the extraordinary wave is heavily affected below an elevation angle of 25°. The difference between ordinary and extraordinary mode absorption also gets much larger at lower elevation angles. Since the range of elevation angles with which the system is concerned, primarily lie below 25°, the need to account for the heightened extraordinary mode losses was further motivated.

Table 4.7 confirms that less absorption was seen approaching the solar minimum. This decrease was apparent even though the calculations of 2005 were done for a lower elevation angle, which would ordinarily increase the amount of absorption taking place. A 1.3 dB difference in ordinary wave absorption at 12:00 UT between 30/12/2001 and 30/12/2005 exists. This is close to the 1.6 dB difference illustrated by ICEPAC. The results of Table 4.7 also demonstrate the fact that more absorption seems to occur in the section of the ionosphere closer to the receiving station at SANAE. This was not only found for December 2005, but for all of the profiles generated. This is likely to be as a result of the effect of the different gyrofrequency and magnetic field inclination at each location.

For the sake of verifying whether the ionospheric model data generated for the Antarctic region is empirical or extrapolated, a manual calculation (not quoted in the results) was done for the absorption of a vertically incident wave at 30 MHz on the date and at the location of a known PCA event. The comparison was done for the study at Terra Nova Bay, where on the 06/11/2001, a PCA event was recorded with a peak vertical 30 MHz absorption of 18 dB at 2:00 UT. Ignoring the gyrofrequency in the calculations, the maximum accumulated absorption using the model data only came to approximately 0.4 dB at 02:00 UT. This was some indication that the data sets output from the models in Antarctica do not necessarily reflect the real ionospheric conditions that occurred. That being said and considering that only quiet time conditions were being accounted for in this analysis, the models used were still sufficient.

4.5.4 Final Ionospheric Absorption Loss Estimates

The peak ionospheric absorption surrounding the solar maximum of 2001 was predicted by ICEPAC to be about 7.76 dB at 12:00 UT in December. This corresponded to an elevation angle of 20.18°. It will be recalled that ICEPAC predicts a lump ionospheric loss which contains non-deviative and deviative absorption losses among others. However, it is not inclusive of the loss seen by the extraordinary mode. In order to extend this quoted value to accommodate the extraordinary mode loss, a supplementary factor needed to be added to the quoted value.

From the set of manual calculations done, the highest level of absorption was seen at 12:00 UT on 20/12/2001. Corresponding to the average elevation angle of 18° used, this gave an ordinary wave absorption of 3.9 dB and an extraordinary wave absorption of 10.87 dB. There was thus a difference of 6.97 dB in the loss of the relative wave modes. This was, however, calculated with respect to a slightly lower elevation angle than that applied in the ICEPAC estimate. Nonetheless, this difference suggests that approximately 7 dB would need to be added to the associated ICEPAC predictions if extraordinary mode absorption is to be incorporated. Of the selected dates and times of solar cycle 23 that were analysed, this means that a worst case quiet ionosphere absorption of roughly 15 dB (7.76 dB + 7 dB) was found. Even though the absorption levels in ordinary conditions are not often likely to be this high, it is important that the system still be able to handle them. This is especially so because it is anticipated that operations will be most reliable in summer months, particularly near to a solar maximum. Even in milder summers such as 2005, ICEPAC predicted a maximum ionospheric loss of 6.16 dB. For the same time, manual calculations yielded 2.38 dB ordinary mode loss and 7.93 dB extraordinary loss. Repeating the same process as above, the relative mode difference of 5.55 dB added to the ICEPAC value of 6.16 dB comes to about 12 dB of total ionospheric loss.

Therefore, assuming quiet ionospheric conditions and accounting for the absorption of the extraordinary mode, an ionospheric loss (L_i) of 15 dB was used in the link budget to ensure system functionality as often as possible.

4.6 Receiver Properties

4.6.1 Receiver Sensitivity

The receiver sensitivity $P_{r,s}$ is the minimum signal power level at which a received signal can be decoded. It can be defined in terms of the signal level required above the total receiver system noise and interference (external and internal), and depends on the individual receiver parameters.

$$P_{r,s} = SNR(dB) + P_{noise}(dBm) \tag{4.5}$$

The SuperDARN receiver has a sensitivity of about $60nV$ [57] with an impedance R of 50Ω which translates to a minimum required receiver signal power $P_{r,s}$ in dBm via,

$$P_{r,s} = 10\log(\frac{V^2}{R \times 1 \times 10^{-3}}) \approx -131[dBm] \tag{4.6}$$

This means that, neglecting the gains of the antennas, a minimum power of -131dBm needs to be received at SANAE in order for the path to be considered open.

The state of the ionosphere is continuously fluctuating and accordingly, so is the amount of power being received. This signal fading can be regular or irregular and has numerous causes e.g. signals arriving via multiple paths, polarisation changes, enhanced absorption and interference [58]. It is difficult to accurately account for all of the losses a signal will experience at all times, which is why a Fade Margin (FM) should always be added to the quoted receiver sensitivity. Including a fade margin of 8 dB and 48 dB, gives a communication reliability of 90% and 99.999% respectively, according to estimates by [59].

ICEPAC includes a fade margin in its output for the total transmission loss. The magnitude of the included fade margin can be estimated from the forecast transmission loss by subtracting free space loss, absorption loss and the transmit and receiver antenna gains specified. It was found that ICEPAC normally only uses a fade margin of between 10-15 dB. For extended reliability, a 35 dB fade margin was the preferred choice for use in the link budget corresponding to $\approx$ 99.95% reliability in accordance with [59].

The minimum power that should be received at SANAE was thus estimated as,

$$P_{r,min} = P_{r,s} + FM = -131 \, [dBm] + 35 \, [dB] = -96 \, [dBm] \tag{4.7}$$

4.6.2 Receiver Antenna

The SuperDARN antennas conventionally have a maximum gain of 21 dBi at 31° elevation according to the characteristics quoted for the Buckland Park radar at 12 MHz [57]. It should be noted that the SANAE radar is installed over a lossy ice shelf ground plane, so the amount of realisable gain is presumably smaller than those quoted at Buckland Park. A receiver antenna gain G_r of 18 dBi was therefore assumed to be more realistic for use in the link budget.

4.7 System Losses

Additional losses from hardware and setup are combined into a system loss denoted L_{sys} which is given by,

$$L_{sys} = L_{ap} + L_{vswr} + L_{eq} \qquad (4.8)$$

L_{ap} is the antenna pointing (AP) loss resulting from the possible misalignment of the transmitter and receiver antennas by an angle (X) and can be found using,

$$L_{ap} = 20 \log \cos X \qquad \text{[dB]} \quad (4.9)$$

Assuming a worst case misalignment of 45° gives this as approximately 3 dB.

L_{vswr} is the VSWR mismatch loss resulting from poor antenna impedance matching and can be found using,

$$L_{vswr} = -10 \log(1 - [\frac{VSWR - 1}{VSWR + 1}]^2) \qquad (4.10)$$

Assuming a worst case VSWR of 2, this results in a 0.5 dB loss. This equates to a 1 dB loss if both the receiver and transmitter antenna are taken into account.

L_{eq} are the losses seen through antenna feed equipment, such as connectors and coaxial cables as well as miscellaneous factors. L_{eq} was estimated as approximately 2.5 dB on each end, equating to a total of 5 dB loss.

4.8 Link Budget Summary

The purpose of the link analysis done was to determine what gain limits the designed antenna should have. This required transmitter antenna gain (G_t) is a function of the transmitter power and can be calculated using,

$$G_t = P_{r,min} - P_t - G_r + L_{fs} + L_i + L_{sys} \qquad (4.11)$$

where,

$$P_{r,min} = P_{r,s} + FM \qquad (4.12)$$

A summary of the link parameters computed can be seen in Table 4.8.

Table 4.8: Summary of link budget parameters for a design frequency of 12.57 MHz.

Link Parameter	Value		
Receiver Sensitivity $(P_{r,s})$	-131 dBm		
Fade Margin (FM)	35 dB		
Minimum Receiver Power $(P_{r,min})$	-96 dBm		
Receiver Antenna Gain (G_r)	18 dBi		
Equipment losses	5 dB		
Free space loss (L_{fs})	122 dB		
Ionospheric absorption loss (L_i)	15 dB		
VSWR mismatch loss	1 dB		
Misalignment loss	3 dB		
Transmitter power (P_t)	**20 dBm (100 mW)**	**27 dBm (500 mW)**	**30 dBm (1000 mW)**
Transmitter Antenna Gain (G_t)	**11 dBi**	**4 dBi**	**1 dBi**

4.8.1 Discussion

Transmitting the minimum power of 100 mW would require a significant antenna gain of 11 dBi, which, given the environmental conditions, was not easy to achieve. Transmitting the maximum power of 1000 mW would be beneficial in that the antenna would only need an antenna gain of 1 dBi. However, given that the aim was to transmit as little power as possible, this was not the best scenario. Therefore, transmitting a power of 500 mW with an antenna gain of 4 dBi was the ideal middle ground that was chosen to design towards.

4.9 Chapter Summary

The key purpose of the path analysis and link budget done was to find the minimum antenna gain needed over the established range of elevation angles so as to ensure a functioning system in normal ionospheric conditions.

The path analysis revealed that elevation angles ranging from $14°$ to $21°$ would usually be needed to obtain an open path. However, dropping the elevation angle lower limit further would be a safer option as at times the predicted angles do in fact fall below $14°$. That being said, lower elevation angles will likely be harder to achieve because of the ground below the antenna.

The link budget done only included estimates of the greatest typical losses encountered during quiet ionospheric conditions along the polar path. This was because ensuring system functionality in periods of appreciable disturbance would require a much higher power to be transmitted as well as a variable operating frequency. It was found that an ionospheric loss of 15 dB should be incorporated so as to consider the propagation of both ordinary and extraordinary wave modes. Hardware and setup losses were added alongside a sizeable fade margin of 35 dB to account for any irregular or unforeseeable losses. From the link budget done, a transmitter power of 500 mW was found to be the most plausible, corresponding to an antenna gain of 4 dBi.

5 Antenna Design and Simulation

5.1 Design Requirements

5.1.1 Technical Specifications

Having investigated more specific system requirements in the previous section, the following transmit antenna criteria were established:

1. Narrowband single operation frequency of 12.57 MHz and horizontal polarisation.

2. VSWR of less than 1.5 ideally (including any additional matching systems required). A VSWR of 2 was, however, accounted for in the link budget.

3. Transmitter power of 500 mW preferably, although a transmitter power up to 1 W can be achieved if necessary.

4. Front lobe gain of at least 4 dBi over the elevation angle range $14° < \theta_{el} < 21°$.

5. Front lobe azimuth beamwidth extending from $-20° < \theta_{az} < 20°$, with at least 4 dBi over the elevation angle range. (This would be a bonus to the system and is not an essential requirement)

6. Maximum back lobe gain directly behind antenna of less than -20 dBi for elevation angles $(\theta_{el}) < 45°$.

7. Maximum back lobe gain of less than -20 dBi within an azimuth angle (θ_{az}) range of $\pm 30°$.

5.1.2 Simulation Optimisation Goal

The primary design goal was to minimise the amount of signal radiated behind the antenna. The amount of backward radiation gain within particular elevation angle ranges was usually reported over actual front-to-back ratios. This made it easier to see and measure what the maximum signal transmitted behind the antenna would be, despite the forward gain of the antenna. The gain and elevation beamwidth were slightly more flexible parameters to adjust. For the horizontally polarised antennas under consideration, these parameters could be varied by altering the mounting height.

Another important aim was to ensure a simple design which will be easy to transport and erect on site. Initially it was thought that size limitations would not matter, bearing in mind that there is plenty of space at the desolate South Pole. It was expected that at a resonant wavelength of 24 metres, the antenna designs would be large. However, the additional complexity of constructing such expansive antennas had not been fully realised. Through the design process, the need to compact the antenna as much as possible became apparent, especially in the vertical plane.

5.2 FEKO Simulation Model and Radiation Pattern Plots

The electromagnetic simulation software used for the antenna designs was FEKO. Simulations were done using the Method of Moments (MoM) integral formulation of Maxwell's equations. For the purpose of preliminary analyses, the conductor material used for all antenna elements was copper. This is because it has the highest conductivity and resistance to corrosion. However, for the final a ntenna d esign, c opper c ladded s teel was chosen based on materials already used by SANSA in the Antarctic environment. The feed frequency (f) was set to 12.57 MHz, which corresponds to a wavelength (λ) of 23.85 m. Feed port input resistance was set to 50 Ω in line with a standard coaxial cable feed. Wire meshing was customised to $1/100\lambda$ as, above this, the accuracy of the formed beam pattern was not found to improve any further. Only far field r adiation p atterns were considered in the analysis done.

An example of the typical polar radiation gain pattern traces that were reported in this book can be seen in Figure 5.1, where a few of the design criteria have been marked. The 3D radiation gain patterns of the same example are shown in Appendix Figure B.2. All polar plot magnitudes were set to decibels relative to an isotropic antenna (dBi).

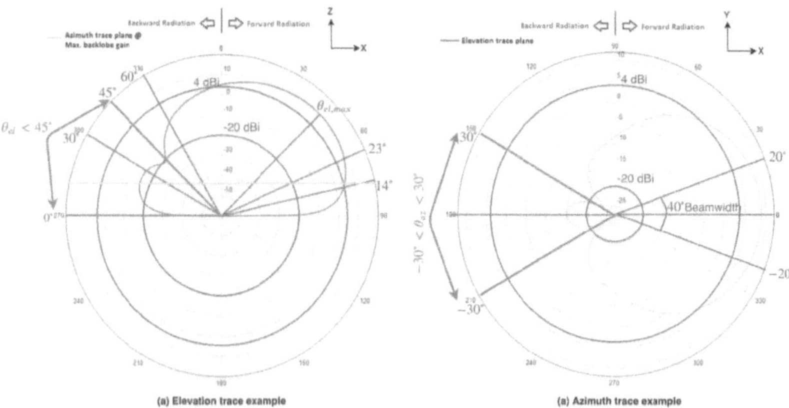

Figure 5.1: Example of the typical radiation gain traces used in the report with some of the technical specifications and limits indicated.

Elevation traces (side views) were depicted for a slice along the xz-plane at the centre of the radiation pattern and antenna (i.e. y=0). Recordings of the maximum back lobe radiation gain were separated into two categories: 1) Elevation angles less than 30° and 2) Elevation angles between 30° and 60°. It should be noted that FEKO displays the elevation angle relative to the vertical axis (z), whereas the elevation angles primarily referred to throughout the design process are relative to the ground axis (x). Hence, the forward elevation angles reported in results were the difference between 90° and the angle displayed on the FEKO polar graphs. Similarly, the backward elevation angles reported were the difference between the displayed angle and 270°. The green lines in Figure 5.1 (a) and (b) indicate the angular limits corresponding to the maximum back lobe gain technical specifications.

Azimuth traces (top views) were shown for a slice along the xy-plane positioned in the z domain at the height of maximum backward radiation below 45° behind the antenna. An example of this is shown by the yellow line in Figure 5.1 (a). This slice of the azimuth was portrayed because the amount of backward radiation at lower angles was an important aspect of the design. On the other hand, half-power azimuth beamwidths ($\Delta\theta_{az}$) were still quoted for a trace sliced at the point of maximum forward radiation ($\theta_{el,max}$). This made the comparison of various antennas easier.

5.3 Ground Plane Model Considerations

5.3.1 Construction of South Pole Ground Dielectric Model

In order to model the ground at the South Pole for use in antenna simulations, the frequency-independent (high frequency) dielectric properties as a function of depth needed to be estimated. The results of this can be seen in Table 5.1 which were obtained as follows:

1. At 10 metre intervals, the density (ρ) and ordinary relative permittivity (ϵ'_r) were approximated using equations (3.6) and (3.7).

2. Using the data in Figure 3.3, approximates of ϵ''_r were extracted at each 10 metre interval.

3. Using the ϵ''_r values in conjunction with equation (3.8) the corresponding high frequency conductivities were calculated.

Table 5.1: Estimated frequency-independent dielectric properties as a function of depth for use in FEKO simulations.

Depth [m]	$\rho\,[kg/m^{-3}]$	ϵ'_r	ϵ''_r @ 200 MHz	$\sigma = \epsilon''_r\epsilon_0\omega\,[S/m]$
0	359	1.7128	0.5×10^{-3}	5.56325×10^{-6}
10	455.5568	1.937048	1.25×10^{-3}	1.39081×10^{-5}
20	535.4053	2.13291	1.625×10^{-3}	1.80806×10^{-5}
30	601.4368	2.302004	1.75×10^{-3}	1.94714×10^{-5}
40	656.0421	2.44671	2×10^{-3}	2.2253×10^{-5}
50	701.1985	2.569707	2.125×10^{-3}	2.36438×10^{-5}
60	738.541	2.6737	2.25×10^{-3}	2.50346×10^{-5}
70	769.4217	2.761256	2.375×10^{-3}	2.64254×10^{-5}
80	794.9588	2.834726	2.5×10^{-3}	2.78163×10^{-5}
90	816.0769	2.896212	2.625×10^{-3}	2.92071×10^{-5}
100	833.5407	2.947557	2.75×10^{-3}	3.05979×10^{-5}

5.3.2 Results of Using Different Ground Plane Models

Using the data in Table 5.1, a simple horizontal half-wave dipole was simulated at 0.25λ high above the following ground plane scenarios:

1. An infinite plane with the linear average of the properties of Table 5.1. I.e. $\rho = 660\,kg/m^{-3}$, $\epsilon'_r = 2.47$ and $\sigma = 2.2\times10^{-5}\,S/m$. (referred to as "average properties")

2. An infinite plane with the surface level properties (At depth=0 m). I.e. $\rho = 359\,kg/m^{-3}$, $\epsilon'_r = 1.7$ and $\sigma = 5.5\times10^{-6}\,S/m$. (referred to as "surface properties")

3. A 100 m multi-layer substrate composed of 10 m layers, each with properties linearly averaged over the respective range segment. These layers are summarised in Table 5.2. (referred to as "multi-layer substrate")

The elevation traces of the simple horizontal dipole mounted 0.25λ above the ground are shown in Figure 5.2. The corresponding magnitude of the antenna characteristic impedances are shown in Table 5.3.

Table 5.2: Multi-layer substrate ground with dielectric properties linearly averaged over 10 m range segments.

Depth [m]	$\rho\,[kg/m^{-3}]$	ϵ'_r	$\sigma\,[S/m]$	Depth [m]	$\rho\,[kg/m^{-3}]$	ϵ'_r	$\sigma\,[S/m]$
0-10	407.3	1.82	9.74×10^{-6}	50-60	719.9	2.62	2.43×10^{-5}
10-20	495.5	2.03	1.6×10^{-5}	60-70	754.0	2.72	2.57×10^{-5}
20-30	568.4	2.22	1.88×10^{-5}	70-80	782.2	2.8	2.71×10^{-5}
30-40	628.7	2.37	2.09×10^{-5}	80-90	805.5	2.87	2.85×10^{-5}
40-50	678.6	2.51	2.29×10^{-5}	90-100	824.8	2.92	2.99×10^{-5}

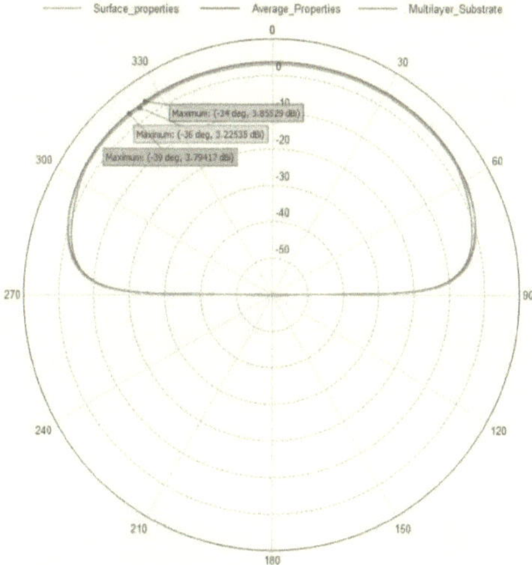

Figure 5.2: Effects of different ground planes on simple horizontal dipole

Table 5.3: Magnitude of the horizontal half-wave dipole characteristic impedance over the various ground planes.

Ground Type	Characteristic Impedance ($|\Omega|$)
Surface properties	115.20
Average properties	114.55
Multi-layer substrate	116.09

5.3.3 Discussion

Figure 5.2 shows that only a slight change in the overall gain of the antenna was seen using different ice ground models for a 12.57 MHz operational frequency ($\approx 0.5\ dBi$

difference). Additionally, just a small decrease of about 3° to 4° in the elevation angle was observed when using the layered ground plane model. These results were not surprising as the antenna is mounted fairly high above the lossy ground. However, this attribute may change depending on the type of antenna being simulated. In order to reduce the computational time taken for initial simulations, the surface properties ground plane was used because they resulted in the least difference in the achieved dipole radiation gain pattern. The dipole's characteristic impedance is also seen to increase by a small amount as the dielectric properties of the ground worsen. Therefore, for simulations of the final antenna, the 100 m multi-layer substrate ground was used in order to predict the antenna overall performance as accurately as possible.

5.4 Terminated Sloping V Antenna Design

5.4.1 Choice of V Antenna over Rhombic Antenna

In order to do a quick comparison of the Rhombic and V antennas, preliminary simulations of terminated wires at 1λ above a PEC plane were done on FEKO for the following configurations:

1. A horizontal terminated V antenna with 4λ elements at an apex angle (θ_{Apx}) of 47.6°.

2. A horizontal terminated Rhombic antenna with 2λ elements at an apex angle of 67.4° (scaled for the shorter individual elements).

A comparison of the resulting *normalised* radiation gain patterns is shown in Figure 5.3. The individual Rhombic and V radiation gain patterns, for both terminated and unterminated wires, can be seen in the Appendix in Figures B.4 and B.3 respectively.

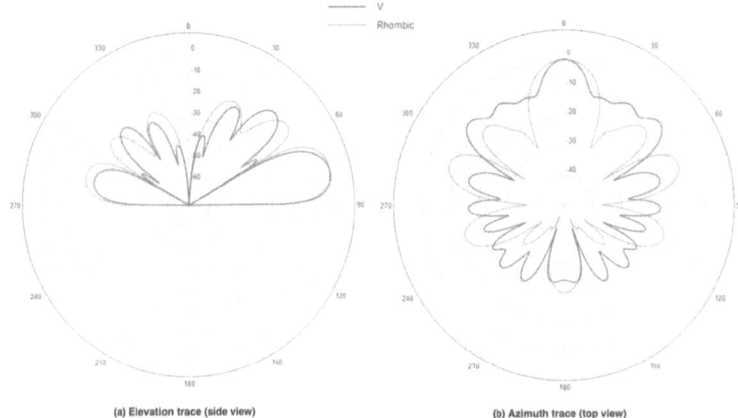

Figure 5.3: Comparison of normalised radiation gain pattern traces for V and Rhombic terminated long wire antenna configurations.

The Rhombic antenna was found to have a cleaner radiation pattern with fewer side lobes and more defined nulls than the V antenna. It also had slightly more front lobe gain ($\approx$ 1 dBi, as expected) and a wider beamwidth. However, as seen in Figure 5.3, the V configuration is actually able to achieve a slightly better main lobe front-to-back ratio ($\approx$5dBi). This advantage is somewhat overshadowed by the more frequent back lobes of the V antenna.

Overall, a Rhombic antenna should perform better than a V antenna. Be that as it may, two of the key antenna design considerations that arose were the ease of instalment and ability to withstand environmental conditions. In either antenna's case, the height above the ground at which the wire elements need to be fed is usually a minimum of 0.5λ, which equates to roughly 12 m high. The V antenna can be sloped, thus requiring only one high mast. On the other hand, the Rhombic antenna would require an additional three masts to suspend all four wire elements in the horizontal plane above the ground. Thus it will be more prone to wind damage and difficult to maintain and erect.

Moreover, achieving as little backward radiation as possible is of high importance. The V antenna can be seen to have 5 dBi less back lobe gain than the Rhombic, which although not particularly significant, is still an advantage. It was therefore decided that the additional complexity of installing a Rhombic antenna outweighed its reported benefits over the V antenna. The design of a terminated sloping V was taken further because of its seemingly simple structure.

5.4.2 Physical Constraints and Simulation Conditions

As discussed earlier on in the literature section, the length of the radiating wire used in a long wire antenna needs to be in the order of wavelengths. The longer the wire used, the higher the achievable gain. The resonant wavelength for which the antenna was being designed was already nearly 24 m long, so the maximum length of wire was limited to 3λ. This is still considerably large, even for a basic physical layout.

Terminated long wire antennas are already particularly inefficient antennas so care needed to be taken to avoid any further unnecessary losses. Having slightly thicker wire gives less resistance, so less power is dissipated through heat. Additionally, a thicker wire will be stronger and less prone to snapping in the environmental conditions. Hence the simulations were done with copper wire of a 0.0002λ diameter. All sloping V antennas were simulated and optimised over an infinite "surface properties" ground plane with $\rho = 359\,kg/m^{-3}$, $\epsilon'_r = 1.7$ and $\sigma = 5.5 \times 10^{-6}\,S/m$.

5.4.3 Design Methodology

The sloping V was the first antenna considered and due to this, practical size limitations had not yet fully been realised. Hence, the sloping V was originally optimised ignoring mast height constraints (referred to as "full-size"). When it became apparent that this size was not feasible, the dimensions were reduced (referred to as "reduced-size").

The next problem that arose was the height at which the wires were being terminated. The length of wire required following on from the terminating resistors was set at 0.25λ as this alleviated the need to terminate in a good ground plane. The initial optimised simulations of the full-size and reduced-size V were done using 0.25λ wire perpendicular to the ground as was seen in Figure 5.4 (a). This, however, cancelled out the benefit of only needing one high support pole (0.25λ is still 6 m high). In an attempt to overcome this, the wire ends were rather terminated in a 0.25λ wire running parallel to the ground as shown in Figure 5.4 (b). In line with adjusting the design to this configuration, a middle-ground termination height (H_t) of 1.5 m was found to still produce minimal back lobe radiation and allow sufficient clearance for snow accumulation. The total height above the ground was limited to 13.43 m ($H_t + 0.5\lambda$). Other than varying leg length and apex angle (θ_{Apx}), all other parameters remained the same as those of the reduced-size V.

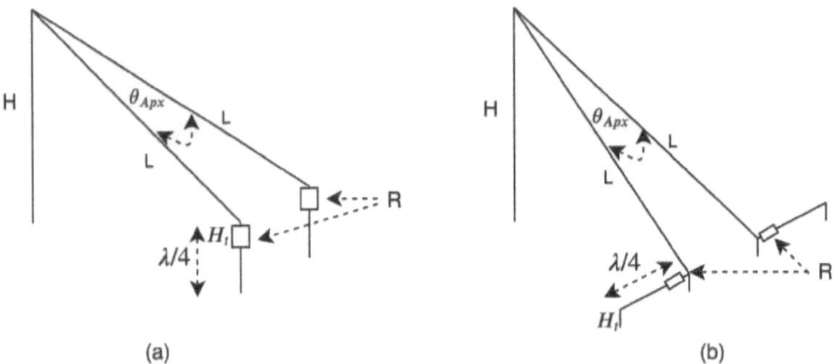

Figure 5.4: Layout variations of the sloping V terminating resistors' height and orientation.

5.4.4   Simulations and Results

Table 5.4 gives a summary of the original optimised full-size and reduced-size sloping V design dimensions found while Figure 5.5 illustrates their radiation gain pattern traces. A few of the key antenna parameters are summarised in Table 5.5 for comparison purposes.

Radiation gain patterns of the adjusted termination height configuration (Figure 5.4 (b)) for different element lengths can be seen in Figure 5.6. Additionally, Table 5.6 was compiled for this setup in order to demonstrate the effect of element length on the amount of achievable gain over the lossy ice ground plane. A 1.5λ leg length was included in these results to demonstrate that elements of anything less than 2λ result in negative gain.

Table 5.4: Dimensions of the optimised full-size and reduced-size sloping V antenna.

Parameter	Full-size	Reduced-size
Leg length (L)	$3\lambda \approx 71.6\ m$	$2\lambda \approx 47.73\ m$
Apex angle (θ_{Apx})	54°	65°
Height above ground (H)	$0.75\lambda \approx 17.9\ m$	$0.5\lambda \approx 11.93\ m$
Termination height (H_t)	$0.25\lambda \approx 5.97\ m$	$0.25\lambda \approx 5.97\ m$
Non-inductive terminating resistors (R)	400 Ω	400 Ω
Wire diameter (w_D)	4.88 mm	4.88 mm

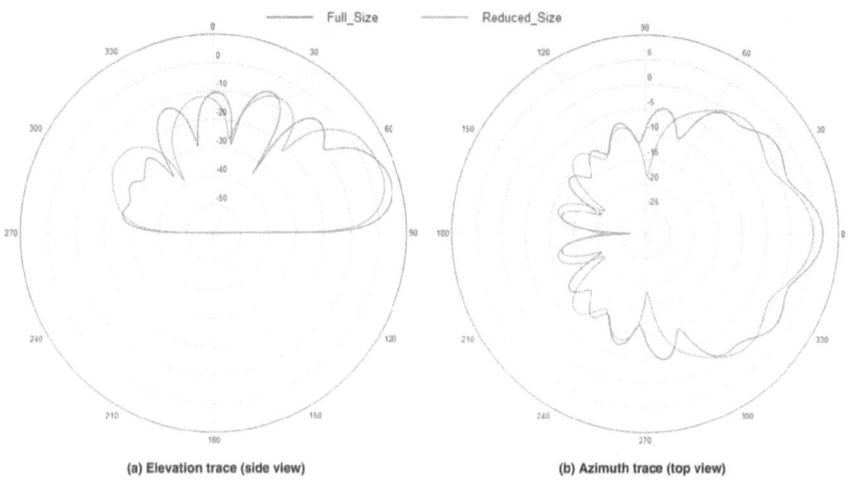

Figure 5.5: Comparison of full-size and reduced-size sloping V optimised radiation gain pattern traces.

Table 5.5: Comparison of measured full-size and reduced-size sloping V radiation parameters.

Antenna type	Max. gain @ $\theta_{el,max}$	Max. backward gain ($0° < \theta_{el} < 30°$)	Max. backward gain ($30° < \theta_{el} < 60°$)	Elevation beamwidth ($-3\ dB$)	Azimuth beamwidth ($-3\ dB$)
Full-size	6.81 dBi @ 16°	$-25.2\ dBi$ @ 30°	$-24.3\ dBi$ @ 47°	16.85°	26.27°
Reduced-size	4.82 dBi @ 20°	$-17.5\ dBi$ @ 30°	$-16.5\ dBi$ @ 35.2°	24.01°	34.64°

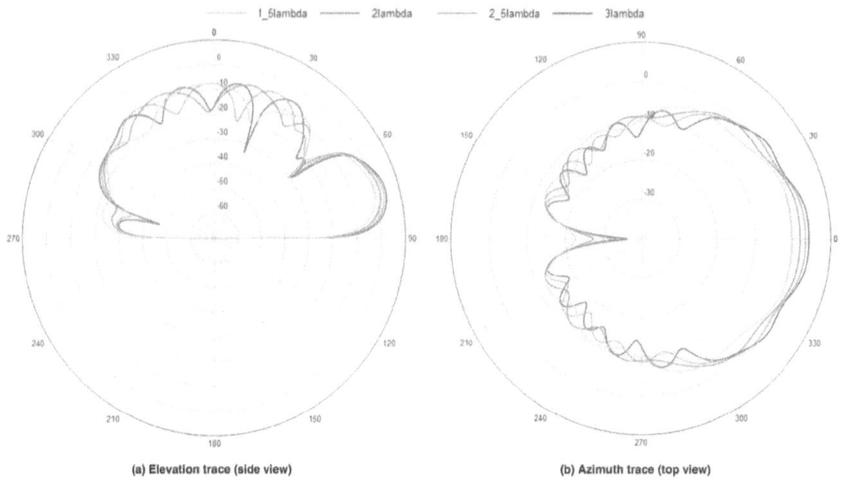

Figure 5.6: Radiation gain pattern traces of the adjusted termination height sloping V configuration for different length elements.

Table 5.6: Gain and elevation angle relation to the element length used in the adjusted termination height sloping V configuration.

Element length	Max. gain @ $\theta_{el,max}$
1.5λ	$-0.33\ dBi$ @ $16.5°$
2λ	$1.99\ dBi$ @ $16°$
2.5λ	$3.40\ dBi$ @ $15.5°$
3λ	$4.35\ dBi$ @ $15.5°$

5.4.5 Discussion

The original full-size sloping V configuration showed promising results, with a front lobe maximum radiation of 6.81 dBi at an ideal elevation angle of 16°. Additionally, anywhere below an elevation angle of 60° behind the antenna, a maximum backward radiation gain of only -24.3 dBi was obtained. This antenna met all of the most important radiation pattern specifications, except that its azimuth beamwidth was not quite broad enough to possibly illuminate other stations. Furthermore, from the azimuth trace, it can be seen that there is only a significant dip in the back lobe gain between $\pm10°$ in azimuth directly behind the antenna. This means that the null behind the antenna would need to be carefully aligned with the South Pole SuperDARN to meet the back lobe radiation requirements. Even by doing so, the null portrayed in simulations will unlikely be obtained in practice. Above all, it was decided that the full-size sloping V dimensions are simply too substantial for real-life implementation in the South Pole environment.

Slightly reducing the size and height of the sloping V still allowed for just enough gain to be supplied over the required elevation angle range. However, the maximum gain of the back lobe radiation below 60° increased to -16.5 dBi, which is above the set limitations. It can be seen from Figure 5.6 and Table 5.6, that the attempts to lower the termination height of the reduced-size V (2λ elements) diminished the maximum front lobe gain to 1.99 dBi. The radiation gain behind the antenna also worsened. To once again increase front lobe gain, 3λ elements would have to be reverted to (giving 4.35 dBi gain). If the maximum power of 1 W was transmitted, shortened elements would no longer be a problem, as only 1 dBi of antenna gain would have to be realised. Nonetheless, the amount of backward radiation achieved with the reduced-size versions of the sloping V was still not quite satisfactory.

5.5 Horizontal Dipole Yagi-Uda Antenna Design

5.5.1 Choice of Yagi-Uda Configurations

Standard horizontal dipole Yagi-Uda's (simply referred to as a "Yagi" from hereon) are amongst the most classic directional antennas and are heavily utilised in industry. They have a simple arrangement and are well-understood. Thus the design of one was undertaken. One of the appeals of the Yagi is that it only requires a single driven element. If multiple elements are to be fed, log-periodic dipole antennas are generally favoured as they produce a far more broadband antenna. However, feeding just two elements as opposed to one does not remarkably complicate construction. For this reason, a dual-feed Yagi design was also considered.

5.5.2 Physical Constraints and Simulation Conditions

Since such a large wavelength is being dealt with, the number of half-wave dipole elements in the Yagi should be kept to a minimum for construction purposes. Element spacing is usually around 0.2λ, so an antenna with anything more than three elements would require a mounting beam length of over 14 m at the wavelength dealt with. Furthermore, establishing a wider beamwidth with a good front-to-back ratio is of a greater priority than adding more director elements to obtain an unnecessarily high gain. Consequently, it was decided to limit all Yagi designs to a maximum mounting beam length of 0.4λ ($\approx$ 10 m). This equates to a limit of three elements for a standard configuration and four elements for a dual-feed. These configurations are revisited in (a) and (b) of Figure 5.7.

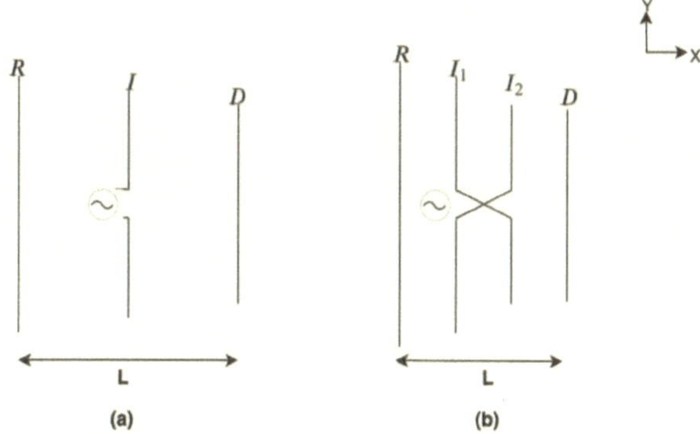

Figure 5.7: Horizontal half-wave dipole Yagi configurations with (a) 3 elements and a single-feed; (b) 4 elements and a dual-feed.

The half-wave dipoles required for a 12.57 MHz operational frequency will span roughly 6 m on either side of the mounting beam. The element material will therefore need to be rigid and so the use of wire is not practical. For the simulations done here, it was decided to model the elements with copper tubing of a standard diameter. It was thought reasonable to set the diameter to around 0.001λ which equates to 23.85 mm. The closest standard copper tube diameter size was 22 mm ($\approx 0.0009\lambda$) and this was what was used for the simulations in FEKO. The actual mounting beam material and size was neglected in light of only preliminary testing being done

5.5.3 Design Methodology

Initial parameters for the three element standard Yagi were found for a 0.001λ diameter based on element lengths recommended by [56]. Basic half-wave dipole Yagi antennas are common and have already been thoroughly investigated in literature, so it was not expected that these parameters would change much. The parameters of a four element dual-feed Yagi with a similar element diameter are not as well-documented and were roughly approximated using recommendations from [53]. The initial antenna parameters for the respective configurations can be seen in Table 5.7.

The design approach taken in this section slightly differed to that of the sloping V antenna, in that the initial parameters were first simulated over a PEC ground. This was done to demonstrate what could be achieved under ideal circumstances. The initial antenna designs were then simulated over an infinite ground plane with $\rho = 359 \, kg/m^{-3}$, $\epsilon'_r = 1.7$ and $\sigma = 5.5 \times 10^{-6} \, S/m$ (simply referred to as "ice ground" from hereon). Optimisation attempts were made for the standard three element Yagi over both the PEC ground and ice ground, but the initial parameters found from literature consistently performed best. However, the four element dual-feed Yagi did need some adjustment for both ground plane scenarios. The antenna mounting height was kept constant for all early Yagi simulations, including the loop element Yagis discussed later on.

Table 5.7: Initial horizontal dipole Yagi parameters adapted from various sources.

Parameter	3 element single-feed	4-element dual-feed
Wavelength (λ)	$23.85 \, m$	$23.85 \, m$
Element diameter (w_D)	$0.001\lambda = 23.85 \, mm$	$0.001\lambda = 23.85 \, mm$
Height above ground (H)	$0.25\lambda = 5.963 \, m$	$0.25\lambda = 5.963 \, m$
Reflector length (R)	$0.507\lambda = 12.092 \, m$	$0.552\lambda = 13.175 \, m$
Driver length (I)	$0.46\lambda = 10.971 \, m$	$I_1 = 0.500\lambda = 11.920 \, m$ $I_2 = 0.433\lambda = 10.332 \, m$
Director length (D)	$0.438\lambda = 10.446 \, m$	$0.4389\lambda = 10.475 \, m$
Element spacing	$0.2\lambda = 4.770 \, m$	$0.1\lambda = 2.385 \, m$
Mounting beam length (L)	$0.4\lambda = 9.54 \, m$	$0.3\lambda = 7.155 \, m$

5.5.4 Simulations and Results

The difference in the radiation pattern of a standard three element Yagi over a PEC ground plane versus over a lossy ice ground plane is exhibited in Figure 5.8. Key antenna parameters are given in Table 5.8. Note that the optimum parameters of the single-feed Yagi ended up being the same as the initial specifications and are mentioned interchangeably. The optimisation process of the dual-feed Yagi can be observed from the traces of Figure 5.9 and the results of Table 5.10. The dimensions established are given in Table 5.9.

Table 5.8: Measured radiation parameters of the 3 element single-feed Yagi over both the PEC and ice ground.

Simulation Conditions	Max. gain @ $\theta_{el,max}$	Max. backward gain $(0° < \theta_{el} < 30°)$	Max. backward gain $(30° < \theta_{el} < 60°)$	Elevation beamwidth $(-3 \, dB)$	Azimuth beamwidth $(-3 \, dB)$
Initial/Optimised PEC ground	$11.12 \, dBi$ @ $44°$	$-15.3 \, dBi$ @ $30°$	$-4.83 \, dBi$ @ $60°$	$52.13°$	$83.47°$
Initial/Optimised Ice ground	$8.15 \, dBi$ @ $35.5°$	$-11.9 \, dBi$ @ $30°$	$-9.62 \, dBi$ @ $60°$	$49.24°$	$73.74°$

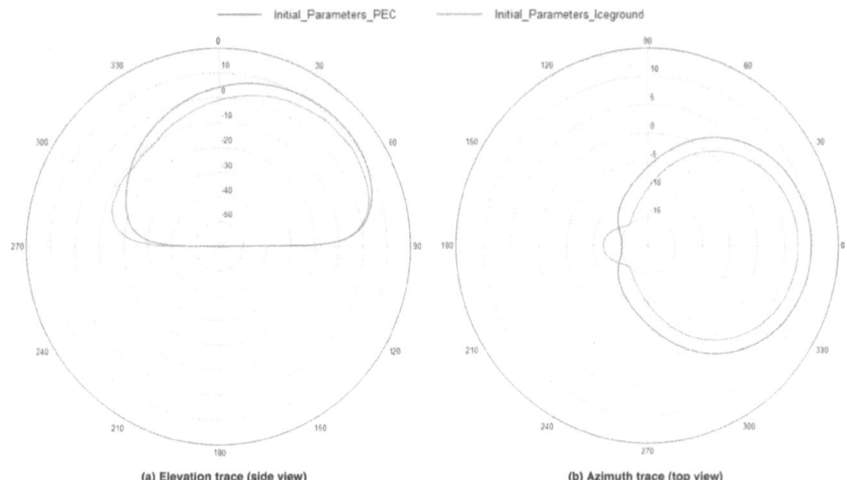

Figure 5.8: Optimised/initial radiation gain pattern traces of the 3 element single-feed Yagi over both the PEC and ice ground.

Table 5.9: Resulting optimised dual-feed horizontal dipole Yagi dimensions over different ground planes.

Parameter	PEC ground	Ice ground
Reflector	$0.55\lambda = 13.126\ m$	$0.544\lambda = 12.983\ m$
Driver(s)	$I_1 = 0.500\lambda = 11.920\ m$ $I_2 = 0.433\lambda = 10.332\ m$	$I_1 = 0.500\lambda = 11.920\ m$ $I_2 = 0.433\lambda = 10.332\ m$
Director	$0.410\lambda = 9.785\ m$	$0.420\lambda = 10.024\ m$
Element spacing	$0.15\lambda = 3.580\ m$	$0.1\lambda = 2.385\ m$
Mounting beam length	$0.45\lambda = 10.740\ m$	$0.3\lambda = 7.155\ m$

5.5.5 Discussion

Antenna Parameter Adjustments

Yagis with only one director generally require uniform element spacing. Independently adjusting the spacing between the respective elements proved futile. The lengths of the driven dipoles have the least impact on the radiation patterns. However, their lengths should be adjusted for resonance once a final layout is settled upon.

The following observations were made when optimising the initial parameters of the four element dual-feed Yagi over a PEC plane:

1. Increasing the element spacing to 0.15λ provided a better front-to-back ratio and pulled down the angle of maximum elevation slightly.

2. Altering the reflector length was not really necessary but 0.55λ was the best by a small margin.

3. Shortening of the director length to 0.41λ decreased the amount of backward radiation to some degree.

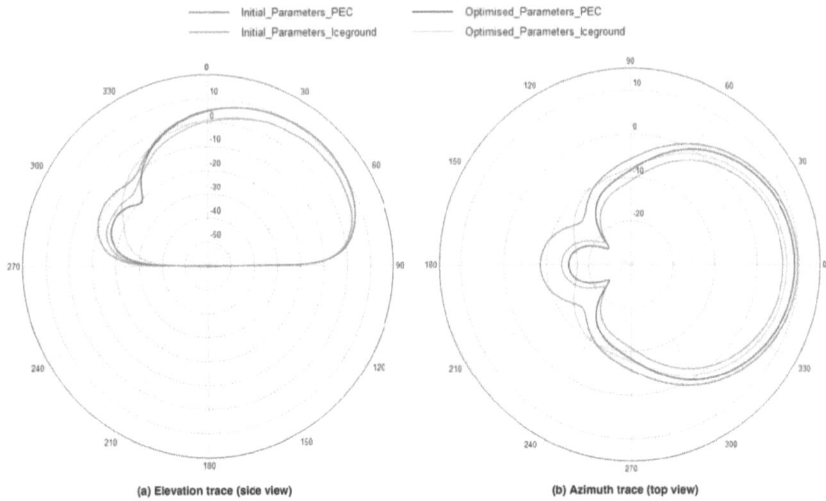

Figure 5.9: Radiation gain pattern traces of the 4 element dual-feed Yagi corresponding to the simulation ground plane used.

Table 5.10: Measured radiation parameters of the 4 element dual-feed Yagi corresponding to the simulation ground plane used.

Simulation Conditions	Max. gain @ $\theta_{el,max}$	Max. backward gain $(0° < \theta_{el} < 30°)$	Max. backward gain $(30° < \theta_{el} < 60°)$	Elevation beamwidth $(-3\,dB)$	Azimuth beamwidth $(-3\,dB)$
Initial parameters PEC ground	10.89 dBi @ 45°	−8.5 dBi @ 28.5°	−6.14 dBi @ 60°	54.91°	86.86°
Initial parameters Ice ground	7.88 dBi @ 35.5°	−13.5 dBi @ 21.4°	−8.86 dBi @ 60°	52.22°	75.57°
Optimised parameters PEC ground	10.99 dBi @ 45.5°	−15.1 dBi @ 24.5°	−6.93 dBi @ 60°	53.95°	87.19°
Optimised parameters Ice ground	7.62 dBi @ 35.5°	−17.4 dBi @ 30°	−5.34 dBi @ 60°	54.5°	77.27°

When optimising initial parameters of the four element dual-feed Yagi over the lossy ice ground plane, it was found that decreasing both the reflector and director length respectively to 0.544λ and 0.42λ improved the overall front-to-back ratio.

On the whole, about 3 dBi less maximum gain was seen when the antennas were mounted over the defined lossy ice ground plane than when mounted over the PEC plane. The optimal 3 element single-feed Yagi parameters were the same for both grounds. On the other hand, the dual-feed Yagi necessitated individually adjusting parameters for each ground plane scenario. This suggests that the dual-feed Yagi may be the less robust antenna of the two, which is likely because it has an extra element that needs to be taken into account.

Performance of Optimised Configurations

A quick comparison of the optimised antennas revealed that the dual-feed Yagi had the smallest maximum backward radiation gain of the two antenna types for elevation angles less than 30°. This effect reversed for elevation angles from 30° to 60°, where it was seen that the single-feed Yagi had less maximum back lobe gain. That being said, neither antenna actually met the defined specifications of less than -20 dBi backward radiation.

The single-feed Yagi had a maximum forward gain of 8.15 dBi, which is an additional 0.5 dBi over the dual-feed for the same elevation angle of 35.5°. In general, this elevation angle is already too high, but could be altered if called for, by adjusting the height above the ground plane. The dual-feed Yagi would in all likelihood be the more preferable of the configurations, if it came down to it. This was decided in view of the fact that the back lobe gain in the range of elevation angles below 30° behind the antenna is more important. The dual-feed Yagi is also smaller in total dimension, despite making use of an extra element.

5.6 Resonant Loop Yagi-Uda Antenna Design

5.6.1 Choice of Loop Shapes

Although not all detailed here, a few different shape elements, namely circular, hexagonal, square and rectangular, were in fact all considered during the initial simulations. It was quickly discovered that the circle and hexagonal shapes performed the most poorly. Furthermore, shaping circular elements for the $\approx 1\lambda$ length perimeters would not be an easy task. Similarly, hexagonal elements would require additional structural components and more careful construction, especially when using wire. In order to keep it simple and based on the fact that most literature tends to point to the use of quadrilateral shapes, only square elements and rectangular elements were selected for further design.

5.6.2 Physical Constraints and Simulation Constraints

Unlike dipole Yagi elements, loop elements are typically made from wire supported by a frame. A medium size copper wire of diameter 0.0002λ was chosen for the initial simulations. Again, the number of elements was limited to three so as to keep the mounting beam length under 10 m. The rectangular element dimensions used were of a 2:1 ratio, with the shorter sides perpendicular to the ground plane. A shorter length of wire in the vertical dimension makes the antenna less susceptible to the ground plane beneath it. This configuration is also beneficial because it reduces the total height of the antenna. The square and rectangular loop configurations can be seen in Figure 5.10 (a) and (b) respectively.

5.6.3 Design Methodology

The initial parameters of a *two* element square loop Yagi were sourced from [53] and a third director element 5% shorter than the active element was added on. The dimensions can be seen in Table 5.11. The same design approach as the horizontal dipole Yagis was taken, namely, the antennas were simulated and optimised at 0.25λ height over both PEC and ice grounds. This time, the antennas were simulated and optimised over the multilayered substrate with properties seen in Table 5.2. This was done because the sides of the loop elements extend closer to the ground than those of horizontal dipoles. The effects of the ground plane medium will once again be touched on in the next chapter. Rectangular Yagi designs are scarce in literature. Therefore the optimised square parameters found

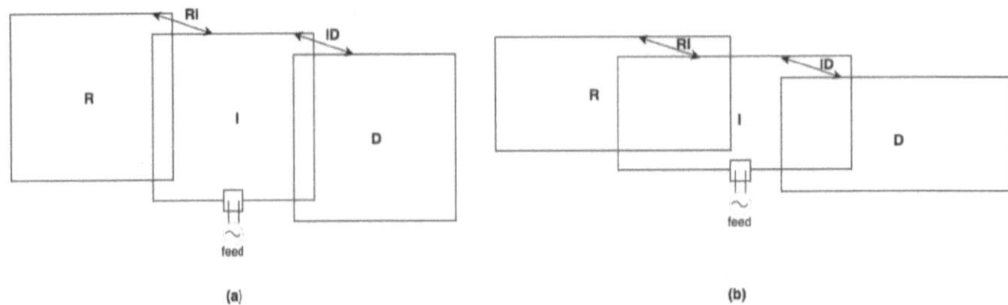

Figure 5.10: Three-element Yagi with:(a) Square loops (b) 2:1 Rectangular loops.

for a PEC plane were used as a basis for the initial rectangular simulations. Note that the different antenna scenarios were sometimes colloquially named in the discussions. For example, the rectangular Yagi over the PEC ground may be referred to as "PEC rectangle", or likewise over the ice ground as "ice rectangle".

Table 5.11: Initial square loop Yagi parameters.

Parameter	Square elements
Wavelength (λ)	23.850 m
Wire diameter (w_D)	$0.0002\lambda = 4.88$ mm
Elements' centre height (H)	$0.25\lambda = 5.963$ m
Reflector perimeter (R)	$1.0757\lambda = 25.673$ m
Driver perimeter (I)	$1.014\lambda = 24.200$ m
Director perimeter (D)	$0.9633\lambda = 22.990$ m
Element spacing	$0.163\lambda = 3.890$ m
Mounting beam length (L)	$0.326\lambda = 8.640$ m

5.6.4 Simulations and Results

The optimised dimensions of both the square and rectangular elements over the ground planes are given in Table 5.12. Figure 5.11 and Figure 5.12 show the acquired radiation pattern traces of the square and rectangular elements respectively. Likewise, recorded key antenna parameters are given in Table 5.13 and Table 5.14. Finally, Figure 5.13 shows a comparison of the optimised ice ground radiation patterns found for both element shapes in order to clearly demonstrate their differences.

Table 5.12: Optimised dimensions of both the square and rectangular loop Yagi antenna corresponding to the simulation ground plane used.

Parameter	Square (PEC ground)	Square (ice ground)	Rectangle (PEC ground)	Rectangle (ice ground)
Reflector Perimeter (R)	$1.07\lambda = 25.537$ m	$1.07\lambda = 25.537$ m	$1.075\lambda = 25.656$ m	$1.07\lambda = 25.537$ m
Driver Perimeter (I)	$1.014\lambda = 24.200$ m	$1.014\lambda = 24.200$ m	$1.014\lambda = 24.200$ m	$1.014\lambda = 24.200$ m
Director Perimeter (D)	$0.925\lambda = 22.076$ m	$0.935\lambda = 22.315$ m	$0.925\lambda = 22.076$ m	$0.9258\lambda = 22.095$ m
Element spacing	$0.163\lambda = 3.890$ m	$0.201\lambda = 4.797$ m	$RI = 0.163\lambda = 3.890$ m $ID = 0.2\lambda = 4.773$ m	$0.175\lambda = 4.177$ m
Mounting beam length (L)	$0.326\lambda = 8.640$ m	$0.402\lambda = 9.594$ m	$0.363\lambda = 8.663$ m	$0.35\lambda = 8.353$ m

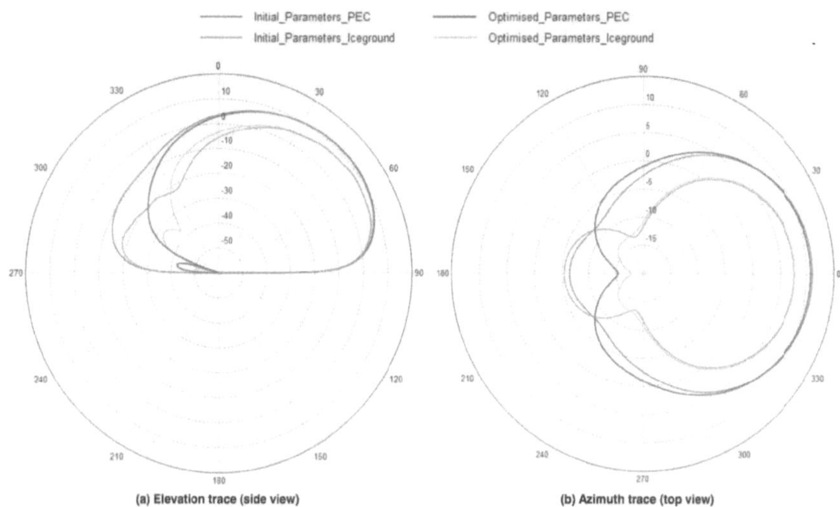

Figure 5.11: Radiation gain pattern traces of the square loop Yagi results corresponding to the simulation ground plane used.

Table 5.13: Measured radiation parameters of the square loop Yagi corresponding to the simulation ground plane used.

Simulation Conditions	Max. gain @ $\theta_{el,max}$	Max. backward gain $(0° < \theta_{el} < 30°)$	Max. backward gain $(30° < \theta_{el} < 60°)$	Elevation beamwidth $(-3\,dB)$	Azimuth beamwidth $(-3\,dB)$
Initial parameters PEC ground	11.12 dBi @ 38.5°	$-6.74\,dBi$ @ 30°	$-6.04\,dBi$ @ 60°	47.96°	79.72°
Initial parameters Ice ground	7.70 dBi @ 34°	$-5.92\,dBi$ @ 30°	$-5.41\,dBi$ @ 35.2°	40.88°	71.97°
Optimised parameters PEC ground	10.80 dBi @ 40°	$-28.6\,dBi$ @ 14.6°	$-4.84\,dBi$ @ 60°	50.44°	84.00°
Optimised parameters Ice ground	7.81 dBi @ 34°	$-20.4\,dBi$ @ 30°	$-11.2\,dBi$ @ 60°	41.48°	73.55°

5.6.5 Discussion

Antenna Parameter Adjustments

When optimising both sets of loop Yagis from the set of initial parameters, the reflector and director perimeters had to be decreased. Director perimeters were found to be closer to 10% smaller than the driver element, as opposed to the 5% commonly expected for a Yagi. The smaller director perimeters appreciably improved the radiation patterns all around. On the other hand, the reflector perimeters required were still fairly close to 5%.

Except for the PEC rectangle Yagi, where a larger spacing between the driver and the director elements (ID) was found most suitable, the Yagi loop elements performed best when evenly spaced. Overall, larger spacing was needed between the elements over the ice ground than the PEC ground and, correspondingly, director perimeters increased a small margin.

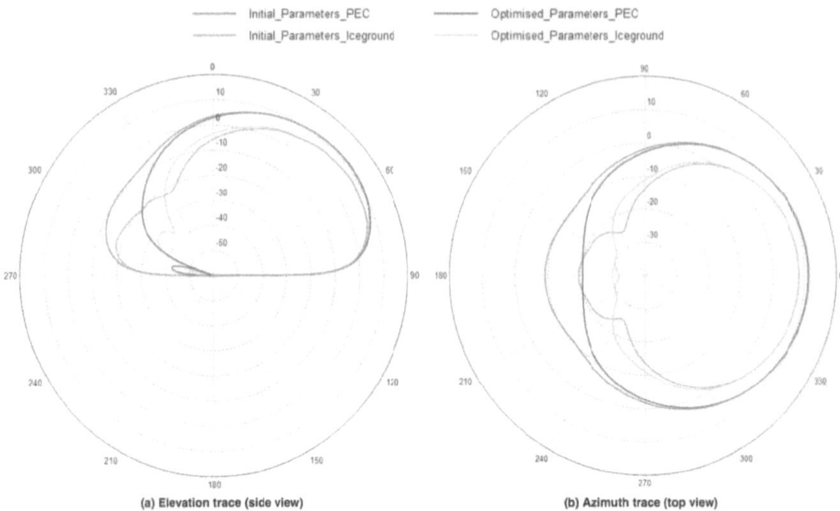

Figure 5.12: Radiation gain pattern traces of the rectangular loop Yagi results corresponding to the simulation ground plane used.

Table 5.14: Measured radiation parameters of the rectangular loop Yagi corresponding to the simulation ground plane used.

Simulation conditions	Max. gain @ $\theta_{el,max}$	Max. backward gain $(0° < \theta_{el} < 30°)$	Max. backward gain $(30° < \theta_{el} < 60°)$	Elevation beamwidth $(-3\,dB)$	Azimuth beamwidth $(-3\,dB)$
Initial parameters PEC ground	11.13 dBi @ 41.7°	−10.4 dBi @ 30°	−6.25 dBi @ 60°	50.78°	81.6491°
Initial parameters Ice ground	8.31 dBi @ 34.5°	−17.7 dBi @ 20°	−20.3 dBi @ 35.2°	41.76°	69.73°
Optimised parameters PEC ground	11.34 dBi @ 41.1°	−29.5 dBi @ 30°	−9.83 dBi @ 60°	49.45°	79.68°
Optimised parameters Ice ground	8.12 dBi @ 34.7°	−33.7 dBi @ 30°	−20.3 dBi @ 60°	43.79°	71.74°

The reflector to driver spacing (RI) was probably the most important when it came to maximising front-to-back ratios and was the same for both the PEC square and rectangle. The main difference between the optimised square and rectangle elements over the ice ground was that the rectangles needed closer spacing and, correspondingly, a smaller director perimeter. It must be recalled that the set of initial parameters used for the rectangle were the same as those found for the square optimised over a PEC plane. The optimised ice rectangle parameters established were in fact very close to these initial parameters, where only an increase of 0.012λ in spacing and 0.0008λ in director perimeter was necessary.

Performance of Optimised Configurations

The final radiation patterns of the antennas over either ground plane quickly demonstrated that the rectangular element Yagi is by far the superior choice of antenna. The maximum ice rectangular Yagi front lobe gain of 8.12 dBi exceeded that of the square

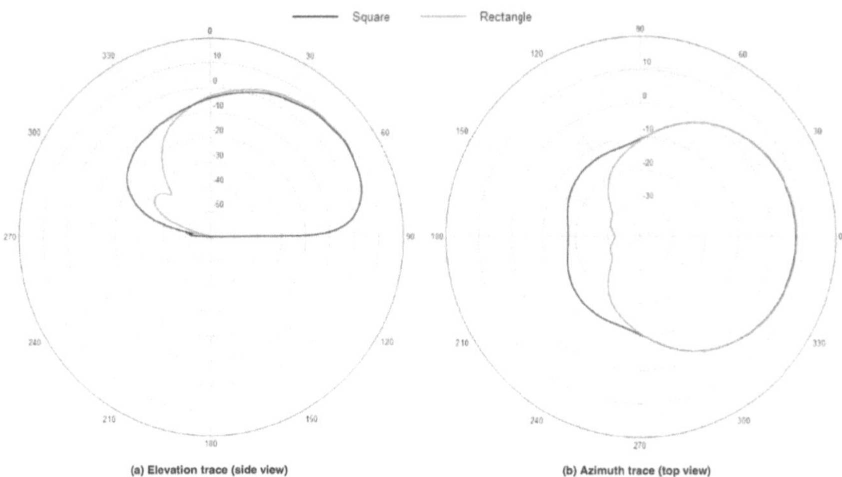

Figure 5.13: Comparison of the optimised radiation gain pattern traces of the square and rectangular loop Yagi over the lossy ice ground plane.

by 0.31 dBi, although it was at an elevation angle 0.7° higher (which is negligible). Additionally, for the same central mounting height of 0.25λ, the vertical dimension of the rectangle would extend to a maximum height of 8.095 m as opposed to 9.159 m for the square.

The rectangular Yagi substantially outperformed the square when it came to minimising the amount of back lobe radiation. Below angles of 30° behind the rectangular Yagi over the ice ground, a maximum of only -33.7 dBi gain was found, while below angles of 60° there was a maximum of -20.3 dBi (and that was at the 60° boundary). If one considers the ice rectangular Yagi azimuth trace in Figure 5.13, taken at the point of maximum back lobe radiation below 45° elevation, it can be seen that, for the whole range of ±60° in azimuth, the back lobe gain lies below -20 dBi. This means that in both azimuth and elevation, the rectangular Yagi meets the maximum back lobe radiation design specifications.

5.7 Summary of Antenna Designs Considered

Table 5.15 shows a summary of the principle results of all the preliminary antenna designs undertaken. All values are cited from the results of optimising the antennas over the "ice ground" plane defined in each scenario.

The sloping V setups were the first antennas properly considered and as a result practical size limitations had not yet been appreciated. They had the potential to perform very well. However, at the system design frequency, the dimensions required to best meet the antenna criteria were simply too large. The smallest length of wire element that would still somewhat align with the technical specifications established was nearly 50 m long. Wires strung out for such lengths would be prone to wind and sagging. Although the winds in the South Pole area are not as extreme as in other regions of Antarctica, this still

Table 5.15: Summary of the measured parameters of the various optimised antenna designs over a lossy ice ground medium.

Antenna type	Max. gain @ $\theta_{el,max}$	Max. backward gain $(0° < \theta_{el} < 30°)$	Max. backward gain $(30° < \theta_{el} < 60°)$	Elevation beamwidth $(-3\,dB)$	Azimuth beamwidth $(-3\,dB)$
Full-size sloping V	6.81 dBi @ 16°	−25.2 dBi @ 30°	−24.3 dBi @ 47°	16.85°	26.27°
Reduced-size sloping V	4.82 dBi @ 20°	−17.5 dBi @ 30°	−16.5 dBi @ 35.2°	24.01°	34.64°
Single-feed dipole Yagi	8.15 dBi @ 35.5°	−11.9 dBi @ 30°	−9.62 dBi @ 60°	49.24°	73.74°
Dual-feed dipole Yagi	7.62 dBi @ 35.5°	−17.4 dBi @ 30°	−5.34 dBi @ 24.5°	54.5°	77.27°
Square loop Yagi	7.81 dBi @ 34°	−20.4 dBi @ 30°	−11.2 dBi @ 60°	41.48°	73.55°
Rectangular loop Yagi	8.12 dBi @ 34.7°	−33.7 dBi @ 30°	−20.3 dBi @ 60°	43.79°	71.74°

leaves the antenna particularly vulnerable to mechanical failures (snapped wires being the most likely). Furthermore, erecting a mast of even 12 m high is still a fairly difficult procedure and may require specialised equipment. Although the sloping V antenna appeared straightforward at first glance, it was found ill-suited to the application.

The principal draw back of the horizontal dipole Yagis once again came down to their physical implementation. The elements, which spanned up to 13 m across, would be tricky to support on a single central mounting beam, even though the elements would be made of rigid material. Furthermore, the performance of the horizontal dipole Yagis was simply not of a high enough standard to justify tackling their mechanical installation complexities.

The loop Yagis were the most compact in total dimension and it was therefore expected that they would be the most viable antennas to erect in the South Pole environment. The rectangular element Yagi not only proved to be the obvious choice between the two loop shapes studied, but also outperformed all preceding antenna designs. It had the least radiation gain in the entire plane behind the antenna, which is important because in most of the other simulated radiation patterns, nulls were only found directly behind the antennas. Relying on achieving such specific nulls is problematic because in real-world implementations they are unlikely to be realised as originally envisaged. Therefore, minimising the total amount of radiation behind the antenna for as broad a range of azimuth as possible was desirable.

Given the analysis done in this chapter, it was decided that a rectangular loop Yagi would be the best transmitter antenna to install at the South Pole. The design of such an antenna is developed further in the next and final design chapter.

6 Final Antenna Design and Analysis

Preliminary antenna simulations indicated that the loop Yagi with rectangular elements would be the most suitable choice. This chapter details the final antenna design considerations of such an antenna. To start off, the impact of changes in wire diameter on optimal parameters were analysed for copper elements. Following which, an appropriate practical element material was selected based on those typically utilised by SANSA and final antenna dimensions were subsequently developed. The physical implementation of the antenna was considered and a brief assessment of the effect of support structure materials on performance was carried out. Finally, the impact of the antenna height above the ground was investigated. A few further tests were also done to establish what margin of accuracy should be involved when constructing the antenna elements.

6.1 Final Considerations

6.1.1 Antenna Resonance

In order for the antenna to be operating at resonance, its characteristic impedance should be purely real (i.e. it should have a reactance of zero). The simulations done in the previous section were not concerned with finding the exact resonant length of the active element, as it had little impact on the radiation pattern for $\pm 0.03\lambda$ of change. For the rectangular loops initially optimised with a copper wire diameter of 4.88 mm, the active element was found to be resonant at a perimeter of 1.015λ and not 1.014λ as was used in the initial analysis. For all subsequent designs, the active element length was always optimised for resonance at 12.57 MHz.

6.1.2 Ground Plane

Besides the loop Yagi antennas, all previous simulations were done over an infinite "surface properties" ground plane with $\rho = 359\,kg/m^{-3}$, $\epsilon_r' = 1.7$ and $\sigma = 5.5 \times 10^{-6}\,S/m$. It was demonstrated that, for a simple horizontal dipole, using a more accurately layered ground had little effect on the resulting performance of the antenna. This result was revisited for the chosen rectangular Yagi, where a comparison was once again done for the antenna over the three different types of ground defined in the earlier sections. The resulting elevation traces are shown in Figure 6.1.

In the case of the rectangular Yagi, it can be observed that the use of a more accurate ground plane did, in fact, alter the radiation pattern of the antenna. This was especially true for the back lobe radiation pattern. The use of different ground planes also impacted the forward gain a small amount (≈ 1 dBi). It was expected that the properties of the snow at the surface would have the most influence on the achievable radiation pattern. This was demonstrated by the fact that the multilayered substrate and the infinite "surface properties" plane produce the most similar radiation traces. All in all, a better radiation pattern was achieved when the antenna was simulated over the infinite "average properties" ground plane. However, the aim was to produce simulations that will reflect real-life performance as closely as possible. Therefore, all subsequent simulations were done over the multilayered substrate of Table 5.2, despite slower computational run time.

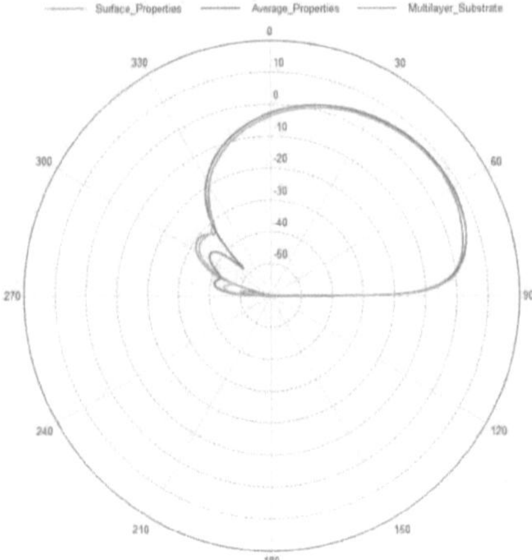

Figure 6.1: The effect of applying different ground planes for simulations of the rectangular element Yagi.

6.2 The Effect of Wire Element Diameter

It was chosen to analyse the changes in parameters required for a wire diameter of both half and double the size of the original 4.88 mm diameter. Copper wire was still employed at this point as in ideal circumstances it would be the best conductor.

6.2.1 Results

The optimised antenna parameter adjustments found can be seen in Table 6.1. Figure 6.2 shows the resulting traces of the radiation gain patterns corresponding to the optimised dimensions for different wire diameters. A few key radiation pattern properties are also summarised in Table 6.2.

Table 6.1: Optimised rectangular Yagi resonant dimensions found for copper wire elements with different diameters.

Wire diameter	Driver perimeter (λ)	Reflector perimeter (λ)	Director perimeter (λ)	Element spacing (λ)
2.44 mm ($\approx 0.0001\lambda$)	1.013	1.065	0.925	0.175
4.88 mm ($\approx 0.0002\lambda$)	1.015	1.07	0.925	0.175
9.76 mm ($\approx 0.0004\lambda$)	1.017	1.075	0.925	0.175

Table 6.2: Measured rectangular Yagi parameters for copper wire elements with different diameters.

Wire radius (mm)	Max. gain and elevation	Max. backward gain $(0° < \theta_{el} < 30°)$	Max. backward gain $(30° < \theta_{el} < 60°)$	Characteristic impedance
2.44	7.33 dBi @ 38.2°	−37.1 dBi @ 30°	−17 dBi @ 60°	121.58 Ω
4.88	7.41 dBi @ 37.9°	−32 dBi @ 30°	−20.5 dBi @ 60°	115.05 Ω
9.76	7.47 dBi @ 37.6°	−27.7 dBi @ 30°	−26.6 dBi @ 60°	110.41 Ω

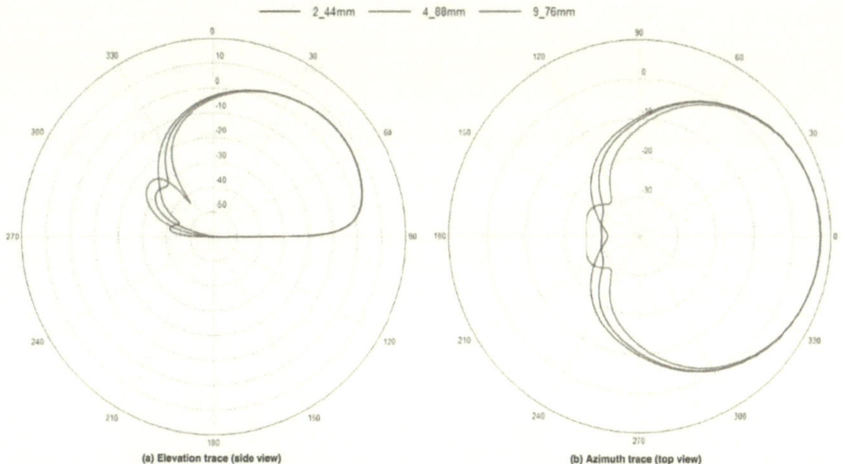

Figure 6.2: Rectangular Yagi optimised radiation gain pattern traces for copper wire elements with different diameters.

6.2.2 Discussion

Antenna Parameter Adjustments

The best overall front-to-back ratio was consistently found at an element spacing of 0.175λ. This optimum spacing was barely impacted by a change in wire diameter and was therefore kept constant. However, decreasing spacing to 0.17λ for the 2.44 mm wire arguably provided a marginally deeper backwards null. Increasing the wire diameter increased the resonant length of the driven element.

The greatest change required was in the length of the reflector. Adjusting it had the most impact on the front-to-back ratio, with an increase in wire diameter corresponding to an increase in element perimeter. The resonant driver and reflector perimeters could be varied independently as they had little effect on one another. The director perimeter required was only slightly affected by the changes in wire diameter. However, for the 9.76 mm wire, decreasing the length to 0.9225λ resulted in about 2 dBi less backward radiation at low elevation angles.

Performance of Optimised Configurations

It was observed that, overall, the smallest diameter (2.44 mm) wire allowed for the least backward radiation, particularly below 30° elevation angles. In the azimuth profile of Figure 6.2, a prominent dip in gain can be observed directly behind the direction of maximum front radiation.

Naturally, a smaller diameter of wire gave a higher characteristic impedance, which resulted in slightly higher antenna feed mismatch loss. Furthermore, approximately 0.15 dBi less gain than the 9.76 mm wire was seen. These additional losses for the smaller diameter wire were, however, considered insignificant. A larger diameter wire would provide a slightly larger bandwidth but a smaller diameter wire was still determined to be more suitable for the design because achieving less backward radiation was preferable.

6.3 Final Antenna Design for Practical Element Material

Copper rod elements such as those used for the horizontal dipole Yagis were considered. However, the additional construction complexity that would likely be required was decided to be unnecessary. The appeal of rod elements lies in their rigidity, making them unlikely to break or move in the presence of wind. Nevertheless, the construction of these elements would require careful welding and making any fine changes to their lengths would be difficult. As mentioned, unlike at the SANAE base station, wind conditions in the South Pole plateau are not particularly severe, with an average wind speed of only 19.8 km/h. It was therefore concluded that wire elements would still be the most practical material as they can be adjusted easily.

The type of wire used was chosen based on its strength and ability to withstand wind and ice. Solid conductors are more likely to break under continual flexing, thus multiple stranded wire would be best. The antenna wire already used by SANSA for the Super-DARN antenna elements is 13 AWG (1.83 mm diameter) 19 strand copper clad steel in a UV resistant Polyethelyne (PE) insulating jacket manufactured by Poly-STEALTH. This has an IACS conductivity of 30% or 40% which translates into 1.74×10^7 $[S/m]$ and 2.32×10^7 $[S/m]$ respectively. The 13 AWG wire is purportedly strong enough for long runs of wire even when lacking additional support [60]. For these reasons, this was the wire chosen for use in the final antenna design.

In FEKO the chosen wire was modelled as a conductor with a 1.83 mm diameter and a conductivity of 2.23×10^7 $[S/m]$ corresponding to a 40% IACS conductivity. A 0.74 mm thick polyethylene insulator coating was applied.

6.3.1 Results

The antenna was roughly re-optimised for the practical wire choice and the resulting final antenna dimensions are shown in Table 6.3. Final radiation traces are seen in Figure 6.3 and recorded antenna parameters shown in Table 6.4. Furthermore, a frequency sweep from 12 MHz to 13 MHz was applied to the optimised model to observe its performance over a wider band. The resulting antenna characteristic impedance plots versus frequency can be seen in Figure 6.4 and the elevation traces of a few select frequencies shown in Figure 6.5.

Table 6.3: Final dimensions of the rectangular element Yagi utilising practical copper-clad steel wire with 1.878 mm diameter.

Parameter	Value
Reflector perimeter (R)	$1.036\lambda = 24.726$ m
Driver perimeter (I)	$0.992\lambda = 23.675$ m
Director perimeter (D)	$0.9185\lambda = 21.921$ m
Element spacing	$0.175\lambda = 4.177$ m
Mounting beam length	$0.35\lambda = 8.353$ m

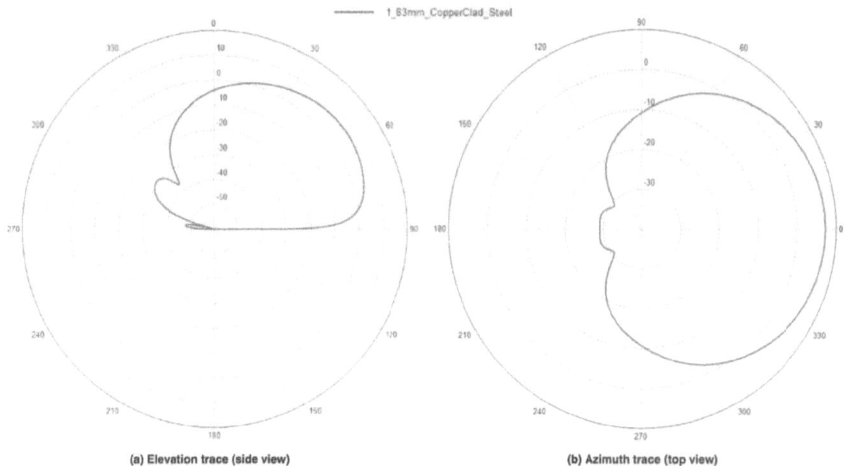

(a) Elevation trace (side view) (b) Azimuth trace (top view)

Figure 6.3: Rectangular Yagi optimised radiation gain pattern traces for practical copper-clad steel wire elements with 1.878 mm diameter.

Table 6.4: Measured rectangular Yagi parameters for practical copper-clad steel wire elements with 1.878 mm diameter.

Parameter	Value
Max. gain and elevation	7.31 dBi @ 37.7°
-3 dBi gain and elevation	4.27 dBi @ 15.7°
Max. backward gain $(0° < \theta_{el} < 30°)$	$-31.7\ dBi$ @ 30°
Max. backward gain $(30° < \theta_{el} < 60°)$	$-23.6\ dBi$ @ 60°
Characteristic impedance	107.55 Ω

6.3.2 Discussion

Understandably, with the smaller dimension of wire considered, all element perimeters were slightly decreased. This time, the resonant perimeter of the driver element fell below one wavelength to 0.992λ. Relative to the resonant driver length, the reflector remained close to 5% longer whilst the director remained close to 7% shorter. Once again, uniform antenna element spacing of 0.175λ provided the best antenna performance.

The final radiation pattern achieved showed that the antenna had well below -20 dBi back lobe radiation for elevation angles less than 60°. Referring to the azimuth trace taken at the slice of maximum back lobe radiation below 45° elevation, a maximum of -29 dBi can be seen within ±30° in azimuth. For ±60° in azimuth, this back lobe radiation still remained less than -20 dBi. In conclusion, this antenna met the desired directional requirements very well.

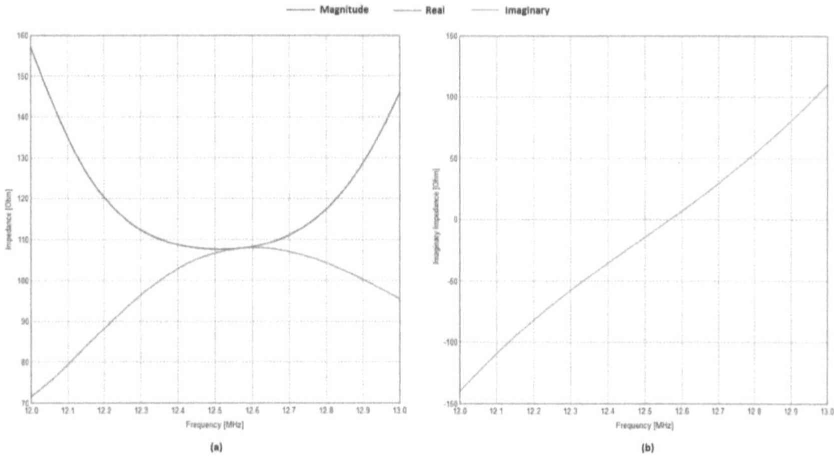

Figure 6.4: Frequency versus the characteristic impedance (a) magnitude and real component; (b) imaginary component.

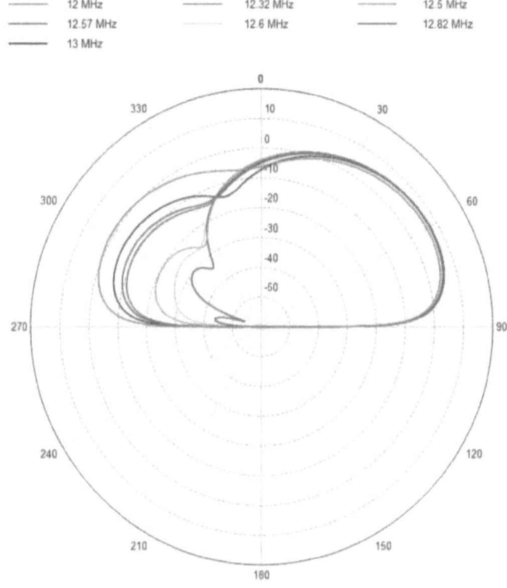

Figure 6.5: Rectangular Yagi radiation gain pattern elevation traces for different frequencies.

It can be seen in Figure 6.4 that the antenna was resonant at 12.57 MHz as, when approaching this frequency, the magnitude of the characteristic impedance becomes equal to the real component, whilst the imaginary component tends to zero. From Figure 6.5 it is apparent that performance of the antenna over the frequency range 12.5 MHz to 12.6 MHz still met the maximum back lobe radiation requirements. However, outside this frequency range, the radiation pattern deteriorated rapidly.

The main problem that still needed to be resolved was that the minimum elevation angle at which 4 dBi gain can be acquired was still a bit too high. Thus, the minimum elevation angle needed to be lowered if only 500 mW power is to be transmitted. These issues are considered in the next section. All ensuing simulations and results were based on the optimised copper-clad steel 1.878 mm diameter wire element antenna.

6.4   Antenna Feed and Matching

The optimised rectangular loop saw an antenna characteristic impedance of approximately 107.55 Ω. Therefore, for an input impedance of 50 Ω, an impedance transformation of around 2:1 will be needed. The VSWR and mismatch coefficients versus frequency can be seen in Figure 6.6. Figure 6.6 (a) shows the results if no matching is applied to the input feed. Figure 6.6 (b) shows the result of increasing the feed impedance to match 100 Ω which would result in a VSWR of 1.08, as opposed to the 2.16 obtained with no matching. The easiest way of implementing this matching would be to use a 2:1 balun such as the one shown in Figure B.6 in the Appendix, which can be procured online from [61].

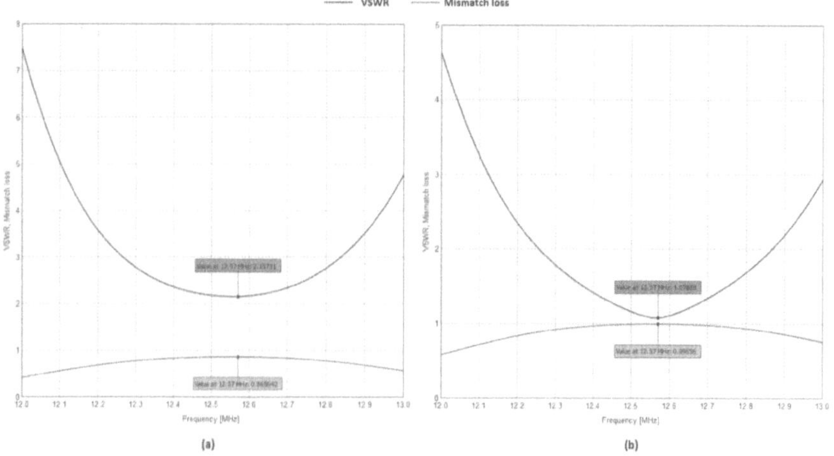

Figure 6.6: Mismatch loss and VSWR parameters for (a) A standard 50 Ω coaxial feed; (b) Matching applied to achieve a 100 Ω input.

A simple adjustable tuning bar can be used at the output of the antenna balun to alter the driven element to exact electrical resonance as the surrounding environment can easily affect this. Furthermore, a short-circuit tuning stub could be attached to the reflector and an open-circuit tuning stub attached to the director, in order to enable adjustment of their reactance on site. This is optional but would nevertheless be a good idea to provide flexibility with assembly and environmental changes.

6.5 Support Structure and Effect on Performance

The rectangular element dimensions were roughly 8 m x 4 m. This made it necessary to design a practical frame to support such a structure. Typically, when wire elements are used in quadrilateral loop Yagi antennas, an X-frame made of non-conductive material is used to shape the elements supported on a central beam. A conductive material such as steel can be used for the central mounting beam and antenna masts, with little effect on the radiation pattern because of the symmetry about the axis. However, the crossed diagonal elements of the frame would need to be nearly 9 m long for the size of rectangular loops being considered. Therefore, keeping all the support structures central would be insufficient mechanically.

A structure was designed as a suggestion to make practical implementation more feasible and is shown in Figure 6.7. This would entail using 4 support masts, 2 central mounting beams and 6 X-frames. In this case, if the mounting beams and masts are made of steel or any conductive material, the antenna performance is significantly affected as the radiating elements are no longer symmetric around them. Specially treated wood would likely be the most suitable material for use in all of the support structures. Plastic materials such as PVC piping are unlikely to withstand the harsh UV exposure and temperatures.

As an alternative to having the radiating wires strung around X frame supports, they could also potentially be laid inside or around a rectangular non-conductive frame. This would protect the elements from blowing in the wind or snapping from the tension around the X frame. It does, however, reintroduce the structural complexity of having to carefully construct a frame fitting to the required element dimensions.

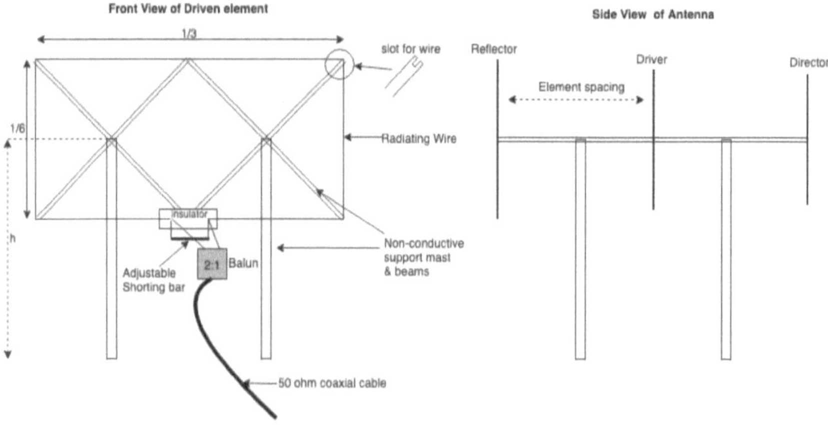

Figure 6.7: Suggested mechanical structure of the final rectangular Yagi.

6.5.1 Results

The effect of using a wooden frame around the elements was simulated on FEKO using
the construction wood dielectric predefined by the program. Slates of wood were placed
in very close proximity to the radiating wires and the result is seen in the elevation trace
of Figure 6.8. Wooden mounting beams and masts were also tested but found to have no
effect on the radiation pattern.

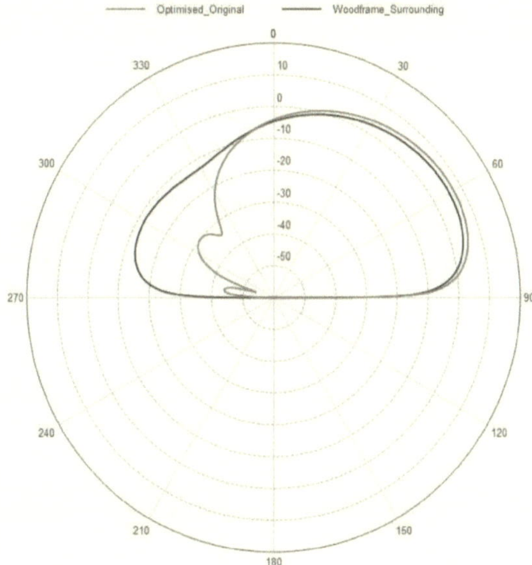

Figure 6.8: Radiation gain pattern elevation traces demonstrating the effect of having a wooden frame
in very close proximity to the radiating elements.

6.5.2 Discussion

As seen in Figure 6.8, the close proximity of the wooden frame to the radiating elements
was found to significantly affect the amount of achievable gain and front-to-back ratio.
It is therefore recommended that the two cross support frames in the same plane as the
rectangular elements suggested in Figure 6.7 are utilised instead as they have no effect on
the radiation pattern.

6.6 Impact of Antenna Height

The optimal antenna parameters that were already established remained consistent independent of the antenna height. However, the radiation pattern achieved will be significantly affected. It was expected that the higher above the ground that the antenna
is mounted, the lower the elevation angle of maximum radiation will be. The maximum
achievable gain will also likely increase with height as the antenna will get further away
from the lossy ground medium.

The performance of the antenna had thus far only been analysed for approximately 6 metres above the ground $(1/4\lambda)$. At this height, the minimum elevation angle that at least 4 dBi gain could be achieved was 14.8°. However, for a 500 mW power supply, it was already found that at least 4 dBi of gain should be realised for an elevation take-off angle ranging from 14° to 22°. Accordingly, either the elevation angle of the designed antenna needed to be lowered slightly or more power needed to be supplied.

In order to find the best mounting height, the final antenna design was simulated over various heights. In this case, the height above the ground refers to the height of the axis around which the rectangular elements are centred. A minimum height of 3 metres was selected so as to allow at least 1 metre space between the ground and the bottom elements of the antenna. This was done to allow for snow accumulation. To keep the total extension of the antenna below 10 metres, a maximum height of 8 metres was set.

6.6.1 Results

The resulting radiation gain patterns of a few selected heights can be seen in Figure 6.9. The resulting half-power gain and corresponding elevation angle are reported in Table 6.5 to show the effect of height on total elevation pattern. The maximum backward radiation gains at elevation angles below 30° are also shown.

It will be recalled that, in the link analysis done, it was found that a transmitter power of 500 mW requires about 4 dBi antenna gain, whereas a transmitter power of 1000 mW only requires about 1 dBi antenna gain. Accordingly for the various heights, the minimum elevation angles for which 1 dBi, 2.5 dBi and 4 dBi can be achieved were reported in Table 6.6. The effect of height on the characteristic impedance is demonstrated by Table 6.7.

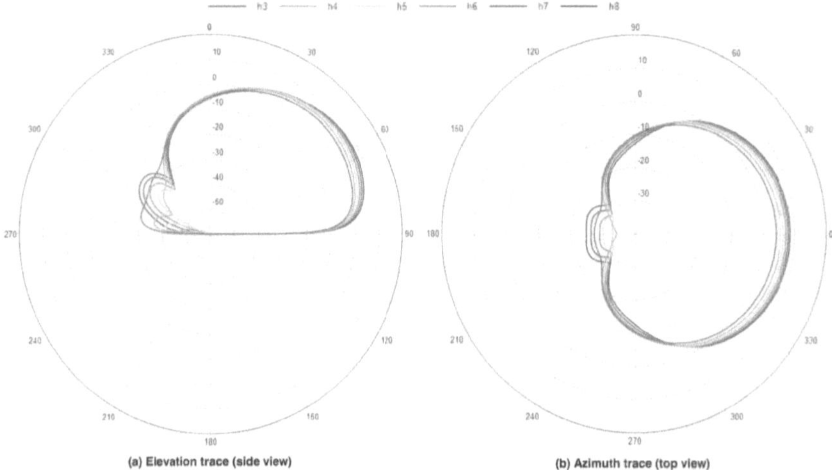

Figure 6.9: Rectangular Yagi radiation gain pattern elevation traces for a range of different antenna mounting heights with the numbers in the key referring to height in m.

Table 6.5: Measured antenna parameters reflecting the impact of adjusting mounting height.

Height (m)	-3 dBi gain and elevation angle	Max. backward gain $(0° < \theta_{el} < 30°)$	Height (m)	-3 dBi gain and elevation angle	Max. backward gain $(0° < \theta_{el} < 30°)$
3	1.03 dBi @ 14.9°	−29.7 dBi @ 30°	6	4.31 dBi @ 15.8°	−31.7 dBi @ 30°
4	2.40 dBi @ 16.2°	−40.8 dBi @ 30°	6.5	4.69 dBi @ 15.3°	−29.7 dBi @ 30°
4.5	2.97 dBi @ 16.3°	−41.2 dBi @ 30°	7	5.02 dBi @ 15.0°	−28.2 dBi @ 30°
5	3.48 dBi @ 16.1°	−37.3 dBi @ 30°	7.5	5.31 dBi @ 14.6°	−27.00 dBi @ 30°
5.5	3.93 dBi @ 16.1°	−34.00 dBi @ 30°	8	5.57 dBi @ 14.1°	−26.10 dBi @ 30°

Table 6.6: Minimum elevation angle at which different gains could be achieved for adjustments to mounting height.

Height (m)	θ_{el} @ 1 dBi (°)	θ_{el} @ 2.5 dBi (°)	θ_{el} @ 4 dBi (°)	Height (m)	θ_{el} @ 1 dBi (°)	θ_{el} @ 2.5 dBi (°)	θ_{el} @ 4 dBi (°)
3	14.9	22.6	41	6	8.6	11.1	14.8
4	12	16.6	24.1	6.5	8	10.2	13.4
4.5	11	14.9	20.7	7	7.5	9.5	12.3
5	10.1	13.2	18.3	7.5	7	8.9	11.4
5.5	9.3	12	16.3	8	6.7	8.3	10.6

Table 6.7: Antenna characteristic impedance relative to height.

Height (m)	Characteristic impedance (Ω)
4	107.52 + j3.46
6	107.55
8	107.13 - j1.41

6.6.2 Discussion

The results of table 6.5 confirmed the expectation that, as the antenna height above the ground increases, the amount of gain will increase and the corresponding elevation angle will decrease. As the height decreased, there was understandably a decrease in the amount of backward radiation due to the smaller overall antenna gain. Nevertheless, as is apparent in the radiation traces of Figure 6.9, the front-to-back ratio itself does actually improve at lower heights. This effect tapers off as the height is decreased below about 4 m and the radiation pattern begins to deteriorate.

Although the minimum required ray take-off elevation angle was set at 14° based on predictions done, it would be worthwhile to drop this even lower to increase reliability. The results of Table 6.6 indicate that, at lower heights, more transmitter power will be needed to compensate for the smaller achievable gain at the required elevation angles. For example, at a height of 4 m, a gain of 4 dBi can only be attained for elevation angles greater than 24.1°. On the other hand, if only 1 dBi gain is required at the same height, elevation angles as low as 12° are still sufficiently supplied with enough gain.

A trade-off was thus needed between having a lower antenna with more transmitter power and having a higher antenna with less transmitter power. Having an antenna mounted as low as possible is convenient for practical reasons but it is not worth the expense of the antenna performance. The fact that more power would need to be supplied at lower heights counteracts the slight improvement in the front-to-back ratio. Snow should also be taken into account as accumulation of it would effectively change the height above the ground and hence the radiation pattern. For this reason there should be a sufficient height margin to ensure the antenna still performs as desired. As can be seen in Table 6.7, the height above the ground does not particularly affect the real component of the characteristic impedance (i.e resistance) but it does alter reactance. The length of the active element would thus need to be adjusted to achieve resonance at different heights.

For consistency, ensuing simulations were still done at a height of 6 m $(1/4\lambda)$ above the ground. However, to ensure that enough gain is delivered to the lower radiation angles and to account for possible snow accumulation below the antenna, it is recommended that this height be set to at least 6.5 m when building the antenna. This would still keep the total height of the antenna below 10 m and would improve overall antenna performance. If the user chooses to mount the antenna lower for convenience sake, 5 m should be the minimum height and at least 700 mW of power should then be supplied. This recommendation is based on the results of Table 6.6, where a power supply of 700 mW corresponds to only requiring about 2.5 dBi antenna gain at the lowest elevation angle.

6.7 Robustness of Final Antenna Dimensions

It is likely that in the construction of the antenna there will be some error in obtaining the exact dimensions recommended flowing from this investigation. For this reason, a parameter length discrepancy of ±10 cm was simulated to observe the impact on antenna performance. In terms of wavelengths, 10 cm equates to approximately $\pm0.004\lambda$ difference in the various parameters. Accordingly, the lengths, spacing and ratio of the rectangular sides, were independently varied by this margin.

The adjusted perimeters of each element can be seen in Table 6.8. In the case of the rectangle side ratios, the adjustments were applied by increasing the horizontal span of all the rectangular elements by 10 cm while decreasing the height proportionately and vice versa (i.e the total length remained the same but the proportions changed).

Table 6.8: Independent parameter adjustments made to the optimised antenna design.

Independently Varied Parameter	Original Length (λ)	Increased Length (λ)	Decreased Length (λ)
Driver Perimeter	0.992	0.988	0.996
Reflector Perimeter	1.036	1.04	1.032
Director Perimeter	0.9185	0.9225	0.9145
Element Spacing	0.175	0.179	0.171

6.7.1 Results

It had already been found that small changes in the length of the driver element mainly affected the reactance of the antenna and not the radiation pattern. Confirmation of this was seen in the elevation trace of Figure B.5 in the Appendix. The consequence of the parameter variations of Table 6.8 can be seen in Figure 6.10 for (a) the reflector perimeter and (b) the director perimeter. The effect of independently varying the rectangular proportions and element spacing can be observed in 6.11 (a) and (b) respectively.

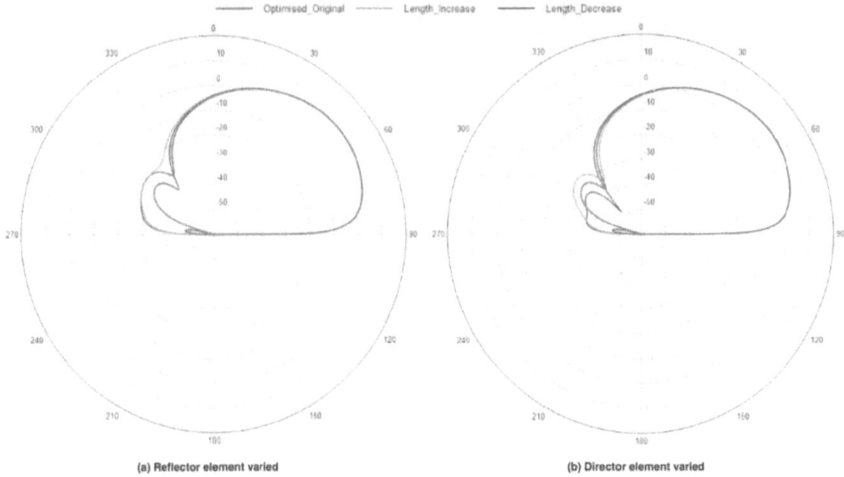

Figure 6.10: Radiation gain pattern elevation traces obtained from individually adjusting the lengths by $\pm 0.004\lambda$ of the rectangular Yagi (a) Reflector; (b) Director.

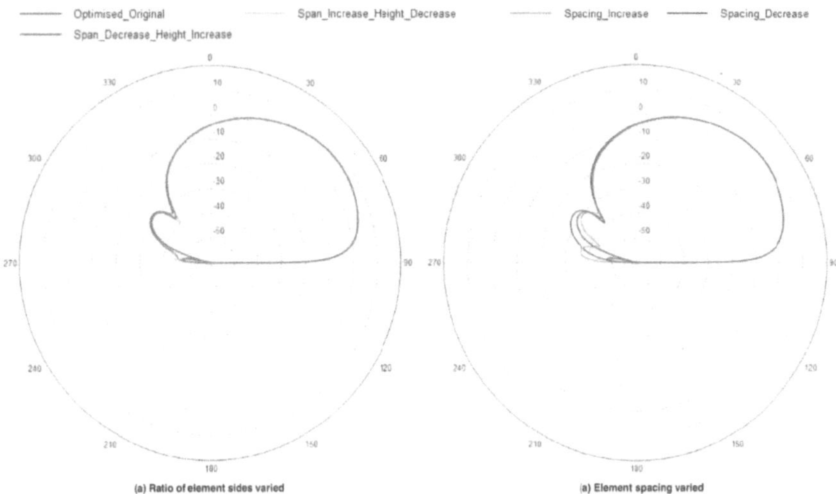

Figure 6.11: Radiation gain pattern elevation traces obtained from individually varying the dimensions by $\pm 0.004\lambda$ of the rectangular Yagi (a) Rectangular proportions; (b) Element spacing.

6.7.2 Discussion

Altering the reflector perimeter was found to have the greatest influence of all on the back lobe radiation pattern, whilst variation of rectangular element proportions (Figure 6.11 (a)) was found to have the least. Nonetheless, for all sets of changes, the amount of back lobe radiation in the elevation plane was still well within the design requirements. This suggests that if small errors are made in the building of the antenna, it will not cause a serious problem. If desired, short-circuit and open-circuit stubs can be added to the reflector and director elements respectively to allow for easy adjustment of their electrical size (refer back to Figure 3.10).

6.8 Final Antenna Summary

6.8.1 Dimensions and Layout

The final rectangular loop Yagi was optimised for elements made of a suitable practical wire available on the market. The reflector was the largest rectangular element, with dimensions of 8.242 m x 4.121 m. The shorter sides of the rectangles were designed to be in the vertical plane, which minimised the total height to which the antenna will extend. The final antenna setup and dimensions are shown in Figure 6.12.

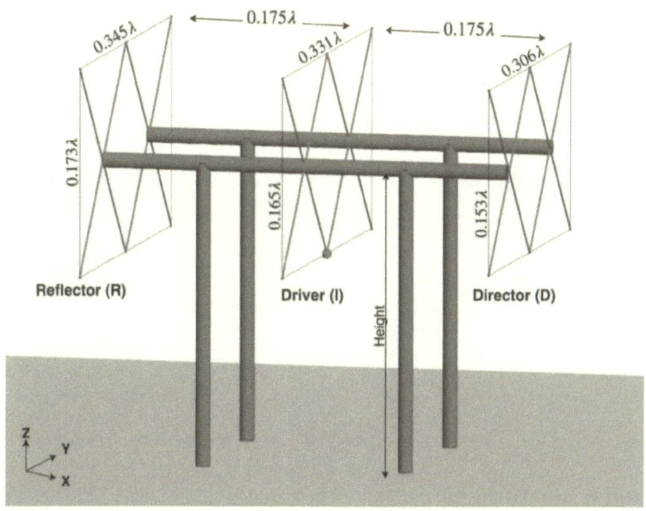

Figure 6.12: Final dimensions and layout of the rectangular Yagi antenna.

6.8.2 Alignment with Technical Specifications

The technical specifications that the antenna needed to meet in order for the design to be deemed a success are recapped here:

1. Narrowband single operation frequency of 12.57 MHz and horizontal polarisation.

2. VSWR of less than 1.5 ideally (including any additional matching systems required). A VSWR of 2 was, however, accounted for in the link budget.

3. Transmitter power of 500 mW preferably, although a transmitter power up to 1000 mW can be achieved if necessary.

4. Front lobe gain of at least 4 dBi over an elevation angle range $14° < \theta_{el} < 21°$.

5. Front lobe azimuth beamwidth extending from $-20° < \theta_{az} < 20°$, with at least 4 dBi over the elevation angle range (This would be a bonus to the system and is not an essential requirement).

6. Maximum back lobe gain directly behind the antenna of less than -20 dBi for elevation angles $(\theta_{el}) < 45°$.

7. Maximum back lobe gain of less than -20 dBi within an azimuth angle (θ_{az}) range of ±30°.

The side and top views of the 3D radiation pattern extracted from FEKO for a central mounting height of 6.5 m is shown in Figure 6.13.

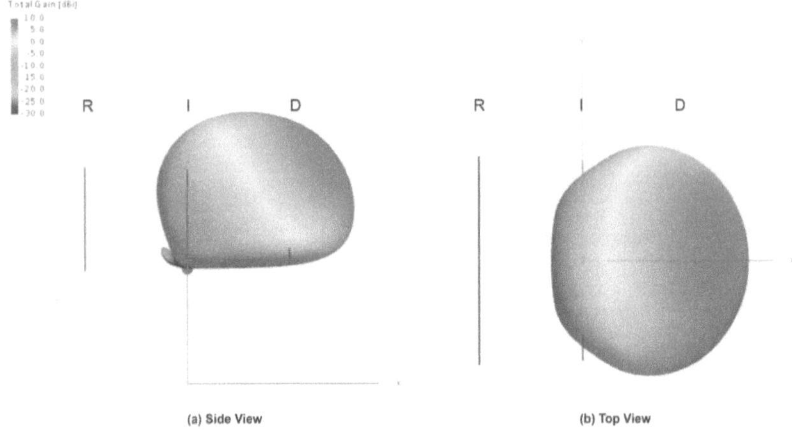

Figure 6.13: Final 3D rectangular Yagi radiation gain pattern simulated on FEKO.

If a 2:1 antenna balun centred on the bottom of the driven element is used to feed the antenna, horizontal polarisation and a VSWR of 1.08 can be obtained. If a non-impedance matching balun is used, a VSWR of around 2.16 will be seen, which would result in 0.63 dB of loss. The link budget analysis that was done already accounted for a worst case VSWR of 2. Thus, even if no impedance matching is applied, it should not be a problem. The first two technical specifications have therefore been met.

A central beam mounting height of at least 6.5 m is recommended in order to supply a gain of 4 dBi extending to a 13.4° elevation take-off angle for a 500 mW power supply. This would also allow for snow build up beneath the antenna. Additionally, at a 13.4° elevation angle, 4 dBi was also achieved over the forward azimuth range of ±20°. This indicated that, overall, the front lobe azimuth beamwidth achieved was broad enough to allow for the simultaneous illumination of Halley Station, which is ≈ 20° apart from SANAE with respect to the South Pole. Syowa Station, which is ≈ 40° apart from SANAE, could also be illuminated if absolutely necessary. However, in order for this to be done, either more transmitter power would have to be supplied or the antenna would have to be aligned with a midpoint between SANAE and Syowa. All of the above results confirm that technical specifications 3, 4 and 5 were met.

At a 6.5 m mounting height, a maximum back lobe radiation of -29.7 dBi was seen behind the antenna below 45° in elevation for an azimuth range of ±30°. Extending the angular limits behind the antenna to 60° in elevation and ±60° in azimuth, a maximum of -20 dBi back lobe radiation was seen. This meant that the performance of the antenna was well within the criteria outlined by technical specification number 6 and 7.

It can thus finally be concluded that the designed 3-element rectangular Yagi successfully met all of the transmitter antenna requirements that were defined for the system under consideration.

7 Conclusions

This project set out to establish the transmitter antenna parameters needed in a 12.57 MHz bistatic ionospheric sounding system for reliable communication over a ≈ 2090 km ground range. This was done, for the most part, by using basic magnetoionic theory in conjunction with available HF propagation tools to conduct an investigation into the ray path that would likely be taken by the transmitted wave in order to reach the receiver. Subsequently, the most notable absorption losses that a radio wave will experience in quiet ionospheric conditions were estimated. A highly directional low-powered HF antenna that met all of the desired specifications was successfully designed. This chapter provides a final overview of the work undergone and briefly answers the set of research questions that were posed in the introduction.

7.1 Answered Research Questions

What suitable ionospheric models and ray tracing methods are easily available for propagation predictions? How valid are these models when applied to the polar regions? What significant shortcomings do they have and how could these potentially be overcome?

The semi-empirical models endorsed by the CCMC are the most widely accepted global ionospheric and geophysical models and were therefore used in analyses. The major problem with such models is that, in the remote Southern Hemisphere high latitudes, data is still fairly scarce, so their reliability is questionable. A quick manual estimate of the absorption experienced by a 30 MHz vertical wave was done using data from the various models for the same date and location of a known and measured polar cap absorption event in Antarctica. The results revealed that the data output from the models at this time did not reflect the disturbed ionospheric conditions that are correlated to an absorption event. This confirmed the concern that the models do not necessarily always represent real ionospheric events in the Southern Hemisphere polar cap region. That being so, it was not expected that the system will be able to operate in highly disturbed conditions in any case. The use of the models for estimating normal ionospheric conditions over Antarctica was therefore considered sufficient.

In terms of actual ray tracers, it was found that, of those freely accessible to all users, the ICEPAC propagation prediction model would be the most suitable. This is because it includes distinct algorithms to account for the effects of a few high-latitude phenomena and can provide estimates of a wide variety of parameters. The online ray tracing tool developed by Virgina Tech for the SuperDARN radars was also utilised because of its ability to retrieve data directly from the IRI model for the exact plane between the South Pole and the SANAE. The main drawback of these ray tracing models is that they fail to take into consideration that the geomagnetic field splits a propagating radio wave into an ordinary and extraordinary mode. ICEPAC and the VT ray tracer only account for the propagation of the ordinary wave and, considering that the SuperDARN will likely receive both modes, this imposes limitations on the predictions made. In order to overcome this to some degree, the transmitter antenna should have sufficient gain to ensure that both the extraordinary and ordinary modes are delivered enough power to reach and be detected by the SuperDARN receiver.

At what times of day and on which dates will the system most likely be able to consistently operate?

Based on predictions made using the VT ray tracing tool, a 12.57 MHz radio wave transmitted from the South Pole will not reach SANAE in years close to, and at, solar minimum. This was concluded based on observations of the solar minimum that occurred in 2008. Insufficient bending of the wave back towards the earth was seen because of the low levels of ionisation. The same effect exists in winter and autumn months. Altogether, the system will function most steadily in November, December and January months (i.e. In summer). As the solar cycle gets closer to solar maximum, the path will be open more frequently. Accordingly, reliable operation can be achieved for the months of October and February in addition to November, December and January.

What is the biggest contributor to ionospheric path loss? What is the peak ionospheric loss that may be experienced by the transmitted signal in standard ionospheric conditions? At what time, season and stage of the solar cycle will this peak occur?

After free space loss, non-deviative absorption that occurs in the D and lower E layer of the ionosphere is the greatest transmission loss. Worst case absorption loss in normal conditions was found to be at 12:00 UT in midsummer at solar maximum. Manual calculations of the non-deviative absorption loss of both extraordinary and ordinary wave modes were done to supplement ICEPAC predictions of total ionospheric loss. Based on the analysis done, it was chosen to include an ionospheric loss of 15 dB in the link budget. The effects of high latitude absorption events were not included in the link because the absorption seen by a 12.57 MHz signal is typically so high that communication is lost completely.

What transmitter antenna gain, power and range of elevation take-off angles is needed to achieve successful communication with the SANAE SuperDARN receiver?

It was found that, ideally, at least 4 dBi of antenna gain should be achieved for elevation angles from 14° to 21° for reliable communication with a transmitter of 500 mW power. The user may choose to rather transmit 1 W of power, thus requiring only 1 dBi antenna gain over the defined range of elevation angles. The key benefit of this is that it would allow for the antenna to be installed closer to the ground.

What are the dielectric properties of the ground plane at the South Pole? What effect does this have on the antenna performance?

Ice and snow generally have much poorer dielectric properties than normal earth or bedrock. Hence, it was already expected that the ice ground at the South Pole would be lossy. It was brought to attention that the earth at the South Pole consists of a layer of firn extending to a depth of about 130 m, before actually developing into the glacial ice that is commonly assumed in antenna simulations over an Antarctic ground plane. Therefore, an attempt was made to model the frequency-independent dielectric properties of a multi-layered substrate so as to create a more accurate FEKO simulation environment. This was done by combining various dielectric relations that have specifically been determined for the South Pole, along with actual data extracted from a study done of an ice core sample that was taken on the South Pole plateau.

It was noticed that the properties of the snow ground at the surface had particular influence on the front-to-back ratio that could be achieved. Accordingly, using a ground plane with dielectric properties that approach solid glacial ice will result in inaccurate expectations of the antenna performance in real conditions. This is especially so when dealing with a highly directional resonant antenna.

What antenna design will meet the developed requirements to the highest degree, whilst being simple to construct, inexpensive and able to withstand the South Pole environment?

Of the antennas reviewed and simulated, a Yagi antenna with rectangular loop elements best met the technical design criteria by an extensive margin. Not only was the achievable radiation pattern the most suitable for the application, but the physical size of the antenna was the most compact at the 23.85 m operational wavelength. Hence, the rectangular Yagi antenna was chosen for the final design.

The final rectangular Yagi was optimised for elements made of a suitable practical wire available on the market. The reflector was the largest rectangular element, with dimensions 8.242 m x 4.121 m. The shorter sides of the rectangles were orientated in the vertical plane so as to reduce the total height to which the antenna would extend. This was important because through the design process undertaken, it was revealed that limiting the total height of the antenna was the most important physical constraint. The rectangular element Yagi can be dismantled easily for transportation to the South Pole, with treated wood likely being the best mechanical support.

A central beam mounting height of at least 6.5 m is recommended in order to supply 4 dBi gain to a $13.4°$ elevation take-off angle and higher for a 500 mW power supply. The user may, however, decide to install the antenna even closer to the ground for practical reasons. In this case, at least 700 mW power should be supplied and the central beam height should not be any lower than 5 m.

In conclusion, a low-powered transmitter antenna has successfully been designed for use in a 12.57 MHz bistatic ionospheric sounding system between the South Pole and SANAE. All of the desired specifications were met to a high standard, as has been discussed in the final antenna summary of Chapter 6.8.2.

8   Recommendations

Lastly, in closing the discussion of the findings detailed in this report, a few recommendations and considerations for possible further investigations are given here.

Antenna Operating Frequency

The fact that the path will largely be closed when approaching a solar minimum is problematic, especially considering there have been speculations that a prolonged solar minimum (coined a "mini ice age") that may last a decade is approaching in 2020 [62] [63]. One recommendation, in order to overcome this problem, would be to transmit a lower frequency wave. In this case, the rectangular Yagi antenna designed would need to be scaled to the new frequency. This could be done easily considering that the dimensions have already been specified in terms of the operational wavelength. However, a lower frequency would result in an even larger antenna. Furthermore, additional antenna gain and power will be required as greater transmission losses are encountered by lower frequency waves. These considerations should be thoroughly investigated before the antenna is built.

Antenna Polarisation

If any other antenna designs are to be considered by the user, it is recommended that the antenna should not necessarily be confined to horizontal polarisation in light of the following reasons:

1. Transmitter antenna polarisation does not need to match receiver antenna polarisation because the radio wave will become elliptically polarised from travelling through the ionosphere [39].

2. There is an already existing SuperDARN antenna of horizontal polarisation stationed at the South Pole. A key requirement was to design a transmitter antenna that interferes with this station as little as possible. Transmitting a vertically polarised wave would mean that any signal radiated directly towards the South Pole SuperDARN would not be picked up by the horizontally polarised system.

3. Structurally wise, some vertically polarised antennas are easier to construct and do not need to be mounted high above the ground. Problems arise because a good antenna ground plane is typically required for the radiation pattern desired to be achieved.

Physical Implementations and Testing

All results reported in this **book** were based purely on simulations. It is there-fore recommended that the antenna be miniaturised by scaling the parameters to a much higher frequency and built for testing in an anechoic chamber facility. A ground material of similar dielectric properties to those found in the South Pole should be used in the chamber if possible. Building and testing of the full size antenna performance over a similar ground range path would also be worthwhile. However, the environmental conditions in South Africa are very different to those found at the South Pole so results may differ notably from those expected.

References

[1] A. Vlasov, K. Kauristie, M. Kamp, J.-P. Luntama, and A. Pogoreltsev, "A study of traveling ionospheric disturbances and atmospheric gravity waves using eiscat svalbard radar ipy-data," in *Annales Geophysicae*, vol. 29, pp. 2101–2116, Copernicus GmbH, 2011.

[2] V. Galushko, A. Kashcheyev, S. Kashcheyev, A. Koloskov, I. Pikulik, Y. Yampolski, V. Litvinov, G. Milinevsky, and S. Rakusa-Suszczewski, "Bistatic hf diagnostics of tids over the antarctic peninsula," *Journal of Atmospheric and Solar-Terrestrial Physics*, vol. 69, no. 4-5, pp. 403–410, 2007.

[3] A. Ribeiro, J. Ruohoniemi, J. Baker, L. Clausen, S. de Larquier, and R. Greenwald, "A new approach for identifying ionospheric backscatter in midlatitude superdarn hf radar observations," *Radio Science*, vol. 46, no. 04, pp. 1–11, 2011.

[4] J. MacDougall, D. Andre, G. Sofko, C.-S. Huang, and A. Koustov, "Travelling ionospheric disturbance properties deduced from super dual auroral radar measurements," in *Annales Geophysicae*, vol. 18, pp. 1550–1559, Springer, 2001.

[5] V. S. Beley, V. G. Galushko, and Y. M. Yampolski, "Traveling ionospheric disturbance diagnostics using hf signal trajectory parameter variations," *Radio Science*, vol. 30, no. 6, pp. 1739–1752, 1995.

[6] G. A. Fabrizio, *High frequency over-the-horizon radar: fundamental principles, signal processing, and practical applications*. McGraw Hill Professional, 2013.

[7] R. D. Hunsucker and J. K. Hargreaves, *The high-latitude ionosphere and its effects on radio propagation*. Cambridge University Press, 2007.

[8] L. F. McNamara, *The ionosphere: communications, surveillance, and direction finding*. Krieger publishing company, 1991.

[9] K. Davies, *Ionospheric radio*. No. 31, IET, 1990.

[10] A. Scott, "Atmospheric waves awareness: An explainer," Apr 2016.

[11] G. R. North, J. A. Pyle, and F. Zhang, *Encyclopedia of atmospheric sciences*, vol. 1. Elsevier, 2014.

[12] M. Voiculescu, I. Virtanen, and T. Nygrén, "The f-region trough: seasonal morphology and relation to interplanetary magnetic field," in *Annales Geophysicae*, vol. 24, pp. 173–185, 2006.

[13] N. Zaalov, E. Moskaleva, and T. Burmakina, "Application of the iri model to the hf propagation model with optimization of the ionosphere parameters to day-to-day variation," *Advances in Space Research*, vol. 60, no. 10, pp. 2252–2267, 2017.

[14] C. Barton, R. Baldwin, D. Barraclough, S. Bushati, M. Chiappini, Y. Cohen, R. Coleman, G. Hulot, P. Kotzé, V. Golovkov, *et al.*, "International geomagnetic reference field, 1995 revision presented by iaga division v, working group 8," *Physics of the earth and planetary interiors*, vol. 97, no. 1-4, pp. 23–26, 1996.

[15] J. Picone, A. Hedin, D. P. Drob, and A. Aikin, "Nrlmsise-00 empirical model of the atmosphere: Statistical comparisons and scientific issues," *Journal of Geophysical Research: Space Physics*, vol. 107, no. A12, pp. SIA–15, 2002.

[16] K. A. Zawdie, D. P. Drob, D. E. Siskind, and C. Coker, "Calculating the absorption of hf radio waves in the ionosphere," *Radio Science*, vol. 52, no. 6, pp. 767–783, 2017.

[17] R. Schunk and A. Nagy, *Ionospheres: physics, plasma physics, and chemistry*. Cambridge university press, 2009.

[18] R. Gillies, G. Hussey, G. Sofko, and H. James, "Relative o-and x-mode transmitted power from superdarn as it relates to the rri instrument on epop.," *Annales Geophysicae (09927689)*, vol. 28, no. 3, 2010.

[19] M. Walden, "The extraordinary wave mode: Neglected in current practical literature for hf nvis communications," 2009.

[20] D. Martyn, "The propagation of medium radio waves in the ionosphere," *Proceedings of the Physical Society*, vol. 47, no. 2, p. 323, 1935.

[21] L. Perrone, L. Alfonsi, V. Romano, and G. De Franceschi, "Polar cap absorption events of november 2001 at terra nova bay, antarctica," in *Annales Geophysicae*, vol. 22, pp. 1633–1648, 2004.

[22] R. Parthasarathy, G. Lerfald, and C. Little, "Derivation of electron-density profiles in the lower ionosphere using radio absorption measurements at multiple frequencies," *Journal of Geophysical Research*, vol. 68, no. 12, pp. 3581–3588, 1963.

[23] J. Patterson, T. Armstrong, C. Laird, D. Detrick, and A. Weatherwax, "Correlation of solar energetic protons and polar cap absorption," *Journal of Geophysical Research: Space Physics*, vol. 106, no. A1, pp. 149–163, 2001.

[24] J. Hargreaves, "Auroral absorption of hf radio waves in the ionosphere: A review of results from the first decade of riometry," *Proceedings of the IEEE*, vol. 57, no. 8, pp. 1348–1373, 1969.

[25] K. G. Budden, *The Propagation of Radio Waves*. Cambridge University Press, 1985.

[26] R. M. Jones and J. J. Stephenson, *A versatile three-dimensional ray tracing computer program for radio waves in the ionosphere*, vol. 1. US Department of Commerce, Office of Telecommunications, 1975.

[27] S. Ritchie and F. Honary, "Advances in ionospheric propagation modelling at high-latitudes," 2009.

[28] D. Bilitza, D. Altadill, V. Truhlik, V. Shubin, I. Galkin, B. Reinisch, and X. Huang, "International reference ionosphere 2016: From ionospheric climate to real-time weather predictions," *Space Weather*, vol. 15, no. 2, pp. 418–429, 2017.

[29] F. Stewart, "Icepac-technical manual," *NTIA/ITS, Boulder*, 2009.

[30] G. Lane, "Review of the high frequency ionospheric communications enhanced profile analysis & circuit (icepac) prediction program," 2005.

[31] "Superdarn ray tracing tool quick guide." http://vt.superdarn.org/tiki-index.php?page=Ray-tracing. Accessed:Aug.28,2019.

[32] C. Coleman, "A ray tracing formulation and its application to some problems in over-the-horizon radar," *Radio Science*, vol. 33, no. 4, pp. 1187–1197, 1998.

[33] *The ionosphere and its effects on radiowave propagation: a guide with background to ITU-R procedures for radioplanners and users.* International Telecommunication Union, 1998.

[34] "Antenna gain, directivity, efficiency." [Online].Available:https://www.elprocus.com/antenna-gain-directivity-efficiency-and-its-conversion/. Accessed:Jun.13,2020.

[35] J. A. Kuch, "Field antenna handbook," tech. rep., Electromagnetic Compatibility Analysis Center Annapolis MD, 1984.

[36] K. Linehan and B. Gianola, "Recommendation on base station antenna standards," *White paper of the NGMN alliance*, 2013.

[37] U. M. Corps, "Antenna handbook," *Washington, DC*, 1999.

[38] R. F. Wireless World, "Rf impedance matching methods - impedance matching devices." [Online].Available:https://www.rfwireless-world.com/Articles/impedance-matching-methods-circuits-devices.html. Accessed:Apr.20,2019.

[39] N. M. Maslin, *HF communications: a systems approach.* CRC Press, 2017.

[40] W. A. Bristow, T. E. Theurer, J. Klein, and M. Guski, "Development of superdarn radars to enable studies of wave polarization," University of Alaska Fairbanks, International Union of Radio Science.

[41] M. C. Rose, M. J. Jarvis, M. A. Clilverd, D. J. Maxfield, and T. J. Rosenberg, "The effect of snow accumulation on imaging riometer performance," *Radio Science*, vol. 35, no. 5, pp. 1143–1153, 2000.

[42] I. Kravchenko, D. Besson, and J. Meyers, "In situ index-of-refraction measurements of the south polar firn with the rice detector," *Journal of Glaciology*, vol. 50, no. 171, pp. 522–532, 2004.

[43] M. van den Broeke, "Depth and density of the antarctic firn layer," *Arctic, Antarctic, and Alpine Research*, vol. 40, no. 2, pp. 432–438, 2008.

[44] V. Schytt, "The inner structure of the ice shelf at maudheim as shown by core drilling," *Norwegian–British–Swedish Antarctic Expedition, 1949–52, Sci. Results*, vol. 4, pp. 113–151, 1958.

[45] N. C. Costes and W. Kingery, "On the process of normal snow densification in an ice cap," *Ice and Snow, Properties, Processes, and Applications*, pp. 412–431, 1963.

[46] O. Eisen, F. Wilhelms, D. Steinhage, and J. Schwander, "Improved method to determine radio-echo sounding reflector depths from ice-core profiles of permittivity and conductivity," *Journal of Glaciology*, vol. 52, no. 177, pp. 299–310, 2006.

[47] O. Eisen, F. Wilhelms, U. Nixdorf, and H. Miller, "Revealing the nature of radar reflections in ice: Dep-based fdtd forward modeling," *Geophysical Research Letters*, vol. 30, no. 5, 2003.

[48] W. D. Callister, D. G. Rethwisch, *et al.*, *Materials Science and Engineering: an introduction*, vol. 7. New York: John Wiley & Sons, 2007.

[49] U. S. N. E. S. Command, *Naval Shore Electronics Criteria: Installation Standards and Practices*. NAVELEX 0101,110, The Command, 1972.

[50] *Recommendation ITU-R BS.705-1: HF transmitting and receiving antennas, characteristics and diagrams*. International Telecommunication Union, 1990-1995.

[51] L. Maes, "Shortwave radio broadcast antennas." https://www.antenna.be/. Accessed: 2019-04-30.

[52] ElectronicsNotes, "Log periodic antenna/aerial." [Online].Available:https://www.electronics-notes.com/articles/antennas-propagation/log-periodic-lpda-antenna/log-periodic-basics.php. Accessed:Apr.30,2019.

[53] T. A. Milligan, *Modern antenna design*. John Wiley & Sons, 2005.

[54] R. Bamford, "The oblique ionospheric sounder," *Project Final Report, Radio Communication Agency*, 2000.

[55] R. D. Straw, *The ARRL antenna book*. 1997.

[56] P. P. Viezbicke, *Yagi antenna design, NBS Technical Note 688*. Washington, DC: US Government Printing Office, 1976.

[57] J. C. Devline and Whittington, "Buckland park radar overview," Jun 2013.

[58] C. J. R. Capela *et al.*, "Protocol of communications for vorsat satellite," 2012.

[59] *The guide to wireless GPS data links*. Pacific Crest, 2000.

[60] "13 awg poly coated poly stealth copper clad steel amateur radio antenna wire." [Online].Available:https://www.k1cra.com. Accessed:Sep.15,2019.

[61] "2:1 hf balun for loop antenna." [Online].Available:https://palomar-engineers.com. Accessed:Oct.10,2019.

[62] N.-A. Mörner, "The approaching new grand solar minimum and little ice age climate conditions," *Natural Science*, vol. 07, pp. 510–518, 01 2015.

[63] "Scientists warn a prolonged solar minimum is coming," *Science Times*, June 2019.

A   HF Link Analysis Auxiliary Material

A.1   VT Ray Tracing Tool Figures

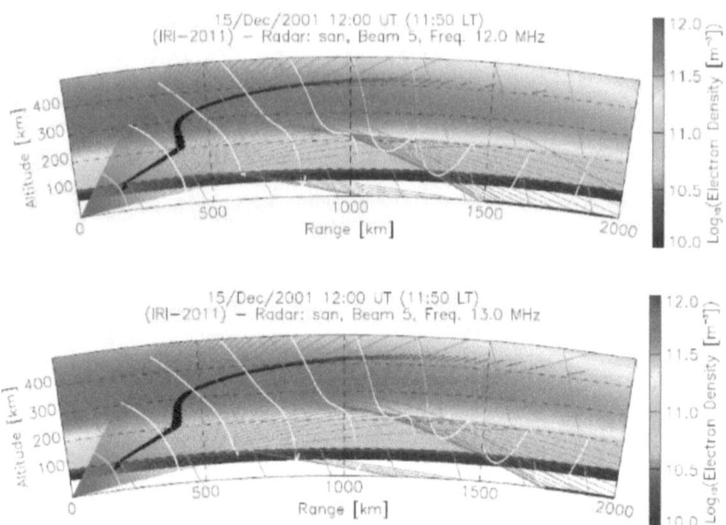

Figure A.1: Differences in the 12 MHZ and 13 MHz SANAE SuperDARN ray traces generated from the VT ray tracing tool.

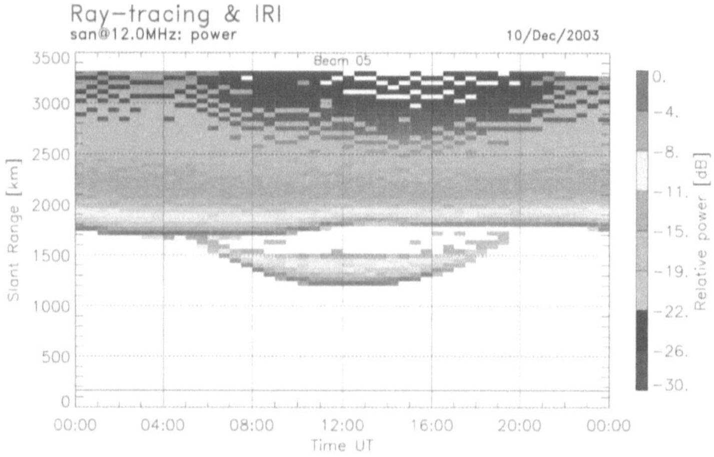

Figure A.2: Example of the power density range plot generated by the VT ray tracing tool.

A.2 Additional ICEPAC Data

```
        JAN   FEB   MAR   APR   MAY   JUNE  JULY  AUG   SEPT  OCT   NOV   DEC

2000   112.9 116.8 119.9 120.8 119.0 118.7 119.8 118.6 116.2 114.6 112.7 112.0
       ( 0)  ( 0)  ( 0)  ( 0)  ( 0)  ( 0)  ( 0)  ( 0)  ( 0)  ( 0)  ( 0)  ( 0)

2001   108.7 104.0 104.8 107.5 108.6 109.8 111.7 113.6 114.1 114.0 115.5 114.6
       ( 0)  ( 0)  ( 0)  ( 0)  ( 0)  ( 0)  ( 0)  ( 0)  ( 0)  ( 0)  ( 0)  ( 0)

2002   113.5 114.6 113.3 110.5 108.8 106.2 102.7  98.7  94.6  90.5  85.2  82.0
       ( 0)  ( 0)  ( 0)  ( 0)  ( 0)  ( 0)  ( 0)  ( 0)  ( 0)  ( 0)  ( 0)  ( 0)

2003    81.0  78.5  74.1  70.3  67.8  65.0  61.8  60.1  59.6  58.1  56.7  54.8
       ( 0)  ( 0)  ( 0)  ( 0)  ( 0)  ( 0)  ( 0)  ( 0)  ( 0)  ( 0)  ( 0)  ( 0)

2004    52.0  49.3  47.1  45.5  43.9  41.7  40.2  39.2  37.5  35.9  35.3  35.2
       ( 0)  ( 0)  ( 0)  ( 0)  ( 0)  ( 0)  ( 0)  ( 0)  ( 0)  ( 0)  ( 0)  ( 0)

2005    34.6  33.9  33.5  31.6  28.9  28.8  29.1  27.4  25.8  25.5  24.9  23.0
       ( 0)  ( 0)  ( 0)  ( 0)  ( 0)  ( 0)  ( 0)  ( 0)  ( 0)  ( 0)  ( 0)  ( 0)
```

Figure A.3: Recommended SNN data for use in the ICEPAC program extracted from the National Geophysical Data Center.

Table A.1: Hourly optimum ray take-off angles predicted by ICEPAC for the path in the solar maximum year of 2001.

2001 Time[UT]	Predicted angle of elevation (°)											
	Jan	Feb	Mar	April	May	June	July	Aug	Sep	Oct	Nov	Dec
2	17.145	11.793	17.225	18.923	18.018	16.713	15.733	16.722	16.681	12.998	16.344	17.419
4	17.766	11.926	14.358	18.169	17.596	16.438	15.266	16.213	16.354	12.777	17.093	17.957
6	18.724	16.292	12.561	17.591	17.065	15.892	14.795	7.667	12.758	12.342	18.193	18.893
8	19.634	16.919	11.755	16.772	16.107	15.158	14.263	15.156	11.171	16.657	18.967	19.534
10	20.141	17.593	14.968	11.968	15.238	14.611	14.028	11.67	10.234	16.817	19.401	19.987
12	20.184	17.738	14.61	10.446	11.591	14.679	13.757	10.391	9.768	16.673	19.573	20.184
14	19.988	17.219	14.191	10.203	10.709	14.644	12.15	9.981	9.492	15.937	19.221	19.897
16	19.333	16.308	10.194	10.209	10.79	14.64	13.741	10.096	9.517	15.322	18.371	19.269
18	18.799	15.549	10.368	10.625	12.411	14.676	13.749	11.804	9.849	14.527	17.558	18.422
20	17.591	10.775	10.896	12.28	15.207	15.156	14.267	15.074	10.657	10.781	16.281	17.737
22	16.983	11	12.992	17.057	16.278	15.652	15.283	16.023	13.178	11.634	15.592	17.077
24	16.664	11.476	17.161	18.181	17.618	16.649	16.127	16.712	16.711	12.609	15.7	17.345

Table A.2: Hourly optimum ray take-off angles predicted by ICEPAC for the path for half of 2005.

2005 Time [UT]	Predicted angle of elevation (°)					
	January	February	March	October	November	December
2	14.849	11.465	14.353	13.367	14.926	14.645
4	15.376	15.911	14.247	13.378	15.143	15.047
6	15.985	15.409	13.916	12.828	15.858	15.543
8	16.601	15.628	11.434	8.103	16.3	15.769
10	17.035	15.932	14.704	8.951	16.543	16.165
12	17.111	16.118	13.817	15.941	16.469	16.16
14	17.065	15.629	13.366	14.868	16.054	15.836
16	16.754	15.185	9.672	14.493	15.654	15.607
18	16.404	14.756	10.106	10.306	15.281	15.308
20	15.625	14.335	10.972	10.887	14.463	14.883
22	14.981	10.855	14.381	13.517	14.294	14.573
24	14.726	11.422	14.893	14.15	14.598	14.65

A.3   Manual Non-Deviative Absorption Calculations

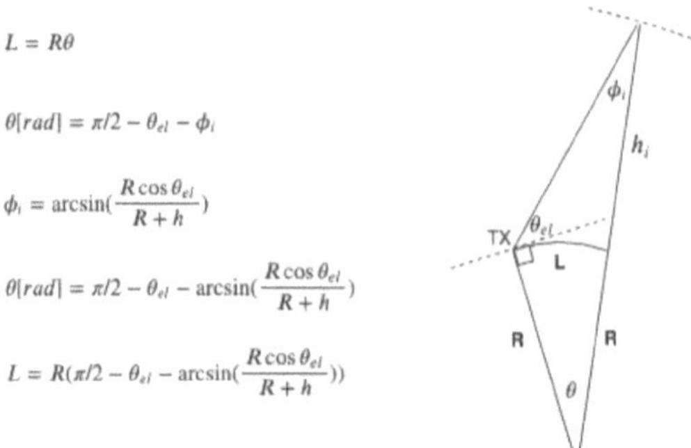

Figure A.4: Geometry and equations used to determine elevation angle, incidence angle and ground range relations.

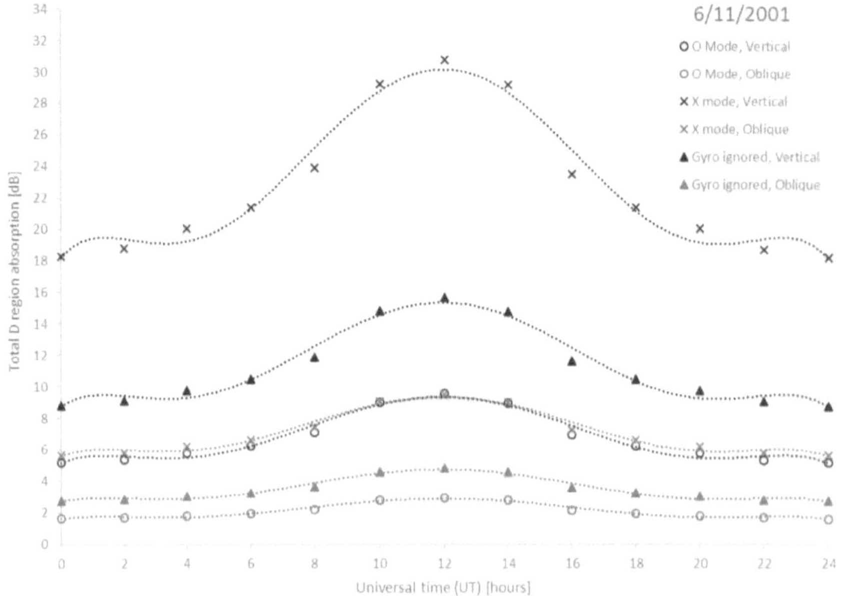

Figure A.5: Hourly non-deviative absorption profiles computed using defined assumptions for November 2001.

B HF Antenna Design Auxiliary Material.

B.1 South Pole Ground Plane

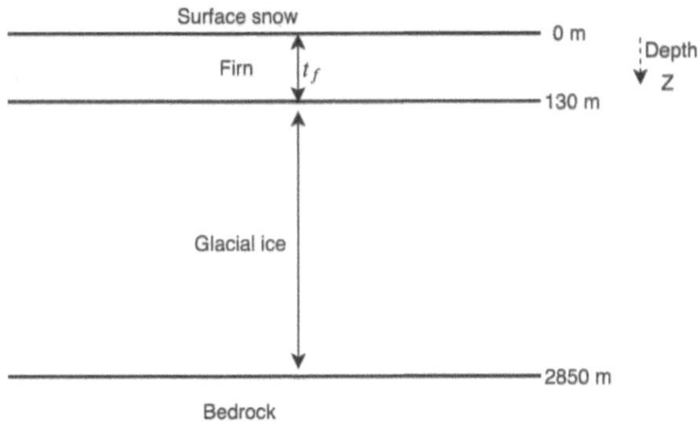

Figure B.1: Schematic of the ground layers at the South Pole.

B.2 FEKO Radiation Pattern Examples

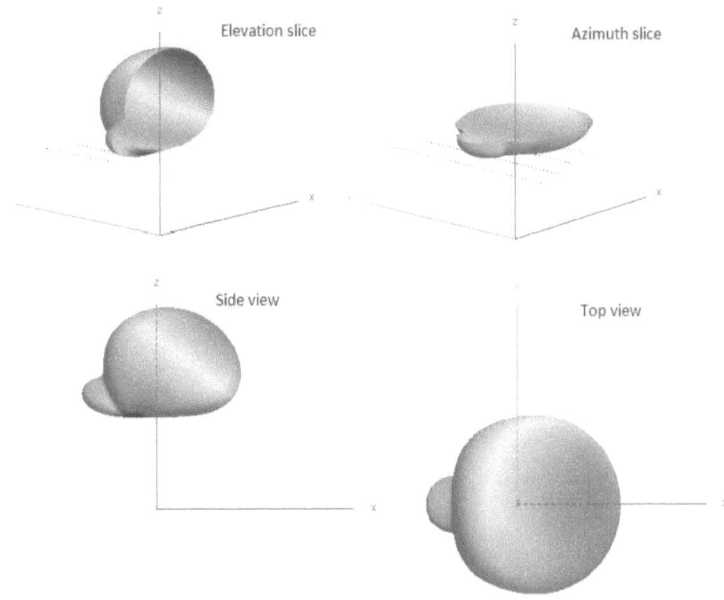

Figure B.2: Example of the 3D FEKO radiation pattern cuts referred to in the report.

B.3 Additional Preliminary Antenna Design Radiation Patterns

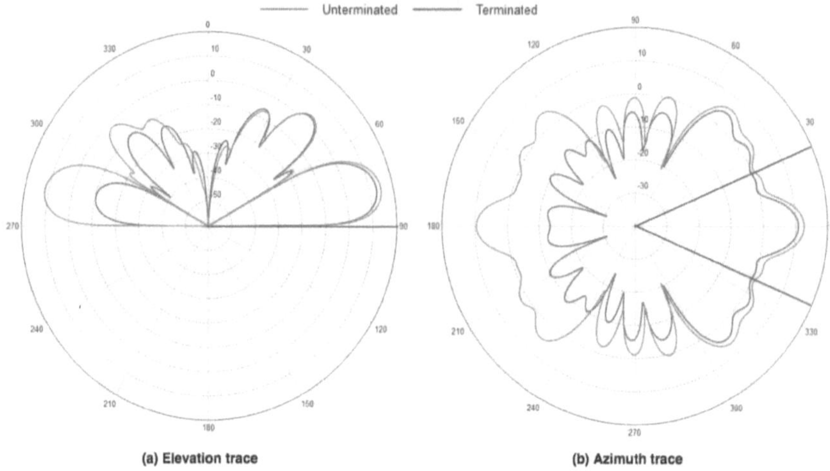

Figure B.3: Resulting radiation pattern traces of both an unterminated and terminated V long wire antenna (position indicated by the blue lines).

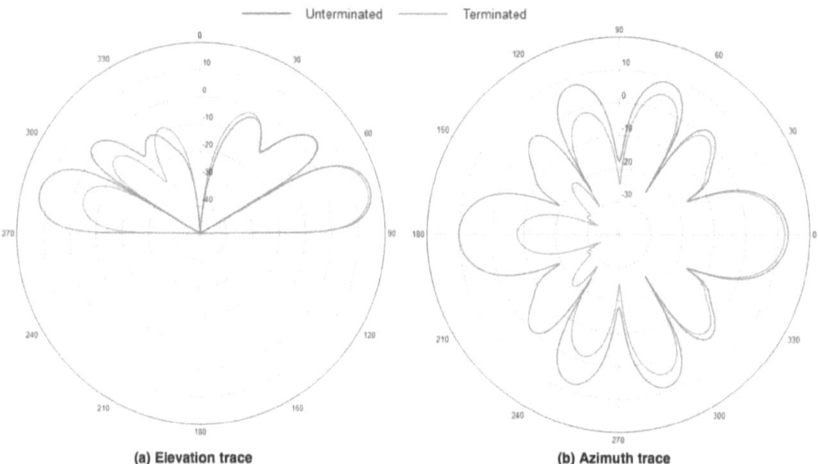

Figure B.4: Resulting radiation pattern traces of both an unterminated and terminated Rhombic long wire antenna.

B.4 Additional Final Antenna Radiation Patterns

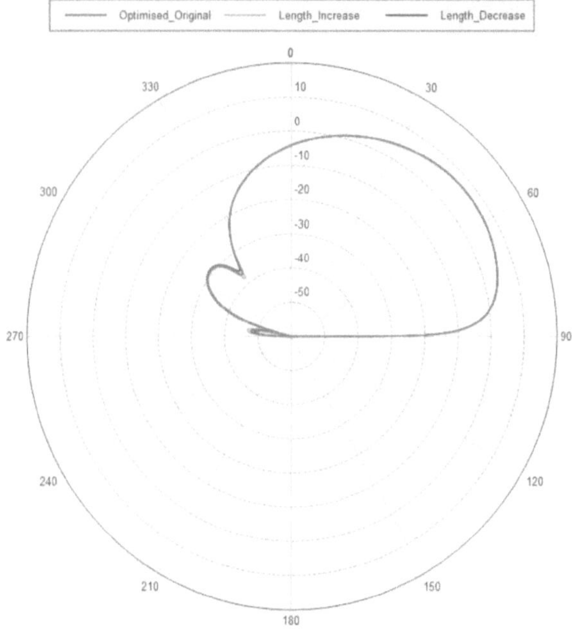

Figure B.5: Effect of varying the length of the active loop element of final rectangle

B.5 Impedance Matching Equipment

Figure B.6: Example of suitable antenna balun secured from Palomar Engineers online at [61]

C Supporting Literature Auxiliary Material

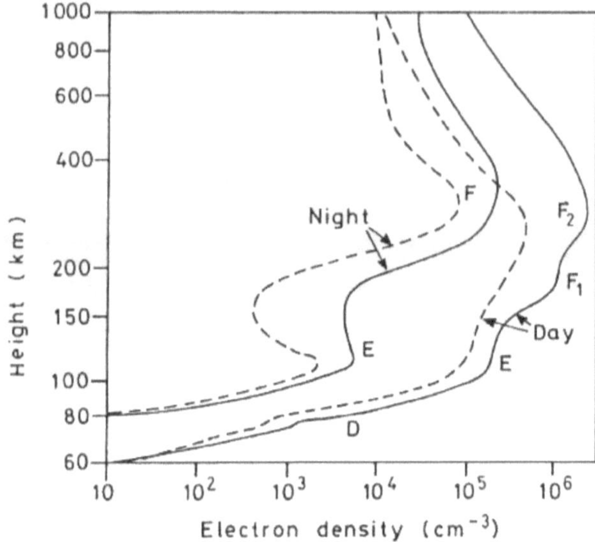

Figure C.1: Example of the typical day/night variations in the vertical electron density profile in a mid-latitude ionosphere extracted from [7]. The solid lines represent a solar maximum whilst the dashed lines represent a solar minimum.

www.ingramcontent.com/pod-product-compliance
Lightning Source LLC
LaVergne TN
LVHW041716190726
843493LV00007B/2112